JOURNAL OF ICT STANDARDIZATION

Volume 2, No. 3 (March 2015)

Special Issue on
Assessments, Models and Evaluation

Guest Editors:
Mr. Malcolm Johnson,
Deputy Secretary-General
International Telecommunication Union

Mr. Chaesub Lee
Director of the Telecommunication
Standardization Bureau (TSB),
International Telecommunication Union

Guest Associate Editor:
Ms. Alessia Magliarditi
Programme Coordinator,
Policy and Technology Watch Division, TSB

JOURNAL OF ICT STANDARDIZATION

Chairperson: Ramjee Prasad, CTIF, Aalborg University, Denmark
Editor-in-Chief: Anand R. Prasad, NEC, Japan
Advisors: Bilel Jamoussi, ITU, Switzerland
Jesper Jerlang, Dansk Standard, Denmark

Editorial Board
Kiritkumar Lathia, Independent ICT Consultant, UK
Hermann Brandt, ETSI, France
Kohei Satoh, ARIB, Japan
Sunghyun Choi, Seoul National University, South Korea
Ashutosh Dutta, AT&T, USA
Alf Zugenmaier, University of Applied Sciences Munich, Germany
Julien Laganier, Luminate Wireless, Inc., USA
John Buford, Avaya, USA
Monique Morrow, Cisco, Switzerland
Vijay K. Gurbani, Alcatel Lucent, USA
Henk J. de Vries, Rotterdam School of Management,
Erasmus University, The Netherlands
Yoichi Maeda, TTC Japan
Debabrata Das, IIIT-Bangalore, India
Signe Annette Bøgh, Dansk Standard, Denmark
Rajarathnam Chandramouli, Stevens Institute of Technology, USA

Objectives
- Bring papers on new developments, innovations and standards to the readers
- Cover pre-development, including technologies with potential of becoming a standard, as well as developed / deployed standards
- Publish on-going work including work with potential of becoming a standard technology
- Publish papers giving explanation of standardization and innovation process and the link between standardization and innovation
- Publish tutorial type papers giving new comers a understanding of standardization and innovation

Aims
- The aim of this journal is to publish standardized as well as related work making "standards" accessible to a wide public – from practitioners to new comers.
- The journal aims at publishing in-depth as well as overview work including papers discussing standardization process and those helping new comers to understand how standards work.

Scope
- Bring up-to-date information regarding standardization in the field of Information and Communication Technology (ICT) covering all protocol layers and technologies in the field.

JOURNAL OF ICT STANDARDIZATION

Volume 2, No. 3 (March 2015)

Published, sold and distributed by:
River Publishers
Niels Jernes Vej 10
9220 Aalborg Ø
Denmark

www.riverpublishers.com

Journal of ICT Standardization is published three times a year. Publication programme, 2014–2015: Volume 2 (3 issues)

ISSN: 2245-800X (Print Version)
ISSN: 2246-0853 (Online Version)
ISBN: 978-87-93237-91-9

Guest Editor Foreword

The Kaleidoscope academic conference is ITU's flagship academic event. Established in 2008, the conference brings the work of ICT researchers to the attention of the standardization community.

The conference has matured into one of the highlights of ITU's calendar of events.

Kaleidoscope sheds light on research at an early stage in the interests of identifying associated standardization needs. Oriented towards the future, the research findings presented to the conference assist ITU in planning the course of its international standardization work.

The sixth edition of Kaleidoscope in 2014 took the theme "Living in a converged world – impossible without standards?". It was tackled from the variety of perspectives that has become essential in the context of technological and industrial convergence.

From a total of ninety-eight submissions from thirty-nine countries, a double-blind, peer-review process selected thirty-four papers for presentation at the conference, all of which were published in the Kaleidoscope Proceedings as well as the IEEE *Xplore* Digital Library.

This is the second in a series of three special issues to showcase extended versions of selected Kaleidoscope papers. The first in the series addressed the theme "Towards 5G" and was published in November 2014 (Volume 2, No.2). The third issue will be published in July 2015.

This second special issue, "Assessments, Models and Evaluation", includes a collection of Kaleidoscope papers that address the performance of ICT networks and emerging capabilities to streamline performance, quality of service (QoS) and quality of experience (QoE). ITU's standardization work targets operational aspects of performance, QoS and QoE; the end-to-end quality aspects of interoperability; and the development of both subjective and objective quality-assessment methodologies for multimedia services. ITU-T Study Group 12 – ITU's standardization expert group responsible for performance, QoS and QoE – is considering the integration of some of these papers' findings into its standardization work.'

The five papers and their respective authors are as follows:

1. Collaboration between Network Players of Information Centric Network: An Engineering-Economic Analysis.
   Mohammad Arifuzzaman (Waseda University, Japan); Keping Yu (Waseda University, Japan); Takuro Sato (Waseda University, Japan)
2. QoXcloud: A cloud platform for QoE evaluation.
   Eduardo Saiz (University of the Basque Country, Spain); Eva Ibarrola (University of the Basque Country, Spain); Leire Cristobo (University of the Basque Country, Spain); Ianire Taboada (University of the Basque Country, Spain)
3. Towards the Standardization of Stereoscopic Video Quality Assessment: An Application for Objective Algorithms.
   José Vinícius de Miranda Cardoso (Federal University of Campina Grande - UFCG, Brazil); Carlos Danilo Regis (IFPB, Brazil); Marcelo S. Alencar (Federal University of Campina Grande, Brazil)
4. On Data Program Interfaces.
   Dmitry Namiot (Moscow State University, Russia); Manfred Sneps-Sneppe (Ventspils University College, Latvia)
5. An Open Source Real-Time Data Portal.
   Sudesh Lutchman (The University of the West Indies, Trinidad and Tobago); Patrick Hosein (University of the West Indies, Trinidad and Tobago)

We would like to thank the authors for their preparation of extended papers, the papers' reviewers for their generous contribution of time and expertise, and of course the readers of this journal for their interest and feedback on this second Kaleidoscope special issue.

Finally, readers are encouraged to participate in the next edition of ITU Kaleidoscope on "Trust in the Information Society". The deadline for submission of full paper proposals is 6 July 2015. Please see the event's webpage: http://itu.int/go/K-2015.

Guest Editors: Mr. Malcolm Johnson
Deputy Secretary-General
International Telecommunication Union

Mr. Chaesub Lee
Director of the Telecommunication Standardization
Bureau (TSB), International Telecommunication Union

Guest Associate Editor: Ms. Alessia Magliarditi
Programme Coordinator,
Policy and Technology Watch
Division, TSB

Collaboration between Network Players of Information Centric Network: An Engineering-Economic Analysis

M. Arifuzzaman*, K. Yu and T. Sato**

Graduate School of Global Information and Telecommunication Studies, Waseda University, Tokyo, Japan
E-mail: {arif; yukeping}@fuji.waseda.jp; t-sato@waseda.jp
*Student Member, IEEE; **Fellow, IEEE*

Received: 14 October 2014;
Accepted: 9 March 2015

Abstract

Internet use is dominated by content distribution and retrieval. As there is a rapid explosion of web-based content, it becomes challenging to provide quality of experience to the end-user. To efficiently handle the utmost growth of the Internet content, in business perspective, there is a clear increase in Content Delivery Networking (CDN). On the other hand, in research perspective it is the Information Centric Networking (ICN) which emerged as an alternative to the current host-to-host communication paradigm and has the potentials of distribution and retrieval of contents more efficiently. This article is an initiative to find some rooms where the ICN and the CDN can meet together. In ICN, a potential strategy by the publisher ignores CDN providers and has direct connections with ISPs. However we believe that, at the stage of incremental deployment of ICN, there is no scope of ignoring CDN. Besides, it is true that it will take time to deploy ICN in Internet scale. So, for achieving big economy of scale publisher will choose CDN as a means for its content distribution. In this paper we present two possible content distribution models for ICN. We formulate game theoretic model for the network players for ICN architecture based on our proposed content distribution model. Firstly, we

Journal of ICT, Vol. 2, 201–222.
doi: 10.13052/jicts2245-800X.231

present two player game model between Publisher and Telco CDN. And we discuss the way of Revenue sharing between Telco CDN and Telco CDN (or Telcos or ISPs); where we assume that interconnection among Telco CDNs is possible. Following, we enhance the game among three major network players (Publisher, Global CDN and Telco CDN) of ICN to analyze the way of the economic incentives sharing among them. We also present a solution for Live Streaming Media broadcast in ICN and analyze the economic part with a decision tree. Besides, we identify some standardization issue in ICN architecture and we emphasize on the need for a common standard for content routers (CR) so that as a node in the ICN, CR ensure scalable content delivery as well as its functionalities match with the Internet open standard philosophy.

Keywords: Information Centric Networking (ICN), Content Delivery Network (CDN), Telco CDN, Telco (Telecommunications Operator), Internet Service Provider (ISP), Quality of Experience (QoE).

1 Introduction

The internet is developing into content network i,e, a video and media network. CDNs play a significant role in the current Internet to optimize the delivery of content. Truly speaking without CDN the current internet would have died already due to lack of capacity. In most of the cases the large content providers use to pay CDNs to deliver their content more efficiently and with guaranteed latencies. ISPs collaborate with CDNs in order to perform such optimized delivery [1]. CDNs proactively push content to servers and then manipulate the Domain Name System (DNS) so as to serve users from nearby servers [2].

On the other hand, Content Centric Network (CCN) [3] is a new paradigm which aims to replace machines by content in the networking communication model. In CCN architecture the network layer provides with content besides providing merely communication channel between hosts. In the PSIRP [4], content-centric networking is used as a base for defining a new architecture. With the Publish-Subscribe internet (PSI) architecture, Information Centric Networking (ICN) focuses on content rather than end point communication. In PSI, any requested information item can be served from the nearby cache or other possible source including replication points. By the nature of the design principle of ICN we can easily recognize that the benefits of ICN dependent on the extensive cache structure in the network. However, it is true that without some explicit monetary gain the network players will not be interested to provide extensive cache structure, neither will it be interested to accommodate a totally new internet architecture. Rather the network player will try to be steadfast with the current internet architecture and continue trying to bridge the gap between the host centric model and the future internet needs by patching to the internet architecture including Mobile IP, CDNs, P2P overlays etc. In this article, we clarify some economic issue of Information centric networking. We recognize the current CDN market scenario to make an engineering economic analysis of ICN. We believe our paper is a good starting point to think the business relationship among different existing network players in the ICN architecture. The major network players we consider are Publisher, Global CDN (example Akamai, Limelight Networks, CDNetworks etc), Telco CDN (CDN developed in an access operator's network i.e., by telecom operators or ISP) and Telcos (or ISPs). The rest of the paper is organized as follows: In Section 2, some related works are discussed. In Section 3, we focus on the standardization issue of ICN architecture. We describe our proposed Content Distribution model for ICN in Section 4. In Section 5, gives the game theoretic model between two player; Publisher and Telco CDN. In Section 6, we enhance the game model between three player; Publisher, Global CDN and Telco CDN and show necessary analysis. In Section 7, we present a solution for Live Streaming Media broadcast in ICN and analyze the economic part with a decision tree. Finally, Section 8 concludes the paper with a summary and some scope of future work.

2 Related Work

Information-Centric Networking (ICN) has gained substantial attention to the research community in recent years as candidate architecture for the future internet. There have been some efforts to address the economic and business aspect of the information centric networking. In [5], the significance of the socio-economic issues in evaluating the future Internet design is clarified and vividly explained. The way of resolving conflicts between the various constellations of stakeholder interests, conflicts etc. are also analyzed in ICN perspective. In [6], the authors present simple economic model to assess the incentives of various network players to establish distributed storage architecture to realize the Information Centric Network. Economic incentives for network players in deploying ICN-based architecture are focused in [7]. Authors also presented a qualitative analysis very precisely. A hierarchical cooperative game model for Resource and revenue sharing among cloud providers has been developed in [8]. Authors have shown that the cooperation and coalition formation can lead to higher profit for the cloud providers. Several technical and business aspects of Content Delivery Network are analyzed from a game theoretic perspective in [9]. For the booming internet-based video consumption, the new revenue opportunity for the Telecommunication and cable providers is unveiled in [10]. In [11], the authors survey the ecosystem for each of the three technical solutions that are proposed in three work packages i.e. Network of Information (NetInf), Open Connectivity Services (OConS) and Cloud Networking (CloNe) of the SAIL (Scalable and Adaptable Internet Solutions) project. Besides the technical architecture, the business architecture of the SAIL project is also precisely presented. In [12], the caching problem in P2P systems is modeled by game theory and authors have shown that the game can reach a Nash Equilibrium. In [13], by using a series of novel payment mechanism, a formal game theoretic model for P2P network is constructed by the authors. The design implications of the case where all users act selfishly to maximize their personal gain are investigated. The paper also analyzed equilibrium of strategies taken by different players. In [14], an analysis of the decision making problem of caching contents by the network players of Information Centric Networking (ICN) is shown. A solution for Live Streaming Media broadcast in ICN and analysis of the economic part with a decision tree is also presented. Besides, the paper also discusses some standardization issue in ICN architecture.

3 Standardization Issue of ICN Architecture

Though the ICN is its infancy, it is candidate architecture for the future internet. Currently, numerous projects are going on under the ICN theme. They vary in their design aspects. We cite here few differences between two major Information Centric Networking project CCN [3] and PSIRP [4] as an example. In case of naming the content, CCN uses hierarchical naming and PSIRP uses flat naming. For security, CCN needs to trust signing key to establish integrity, where in case of PSIRP it is self- certifying. For name resolution and routing, name based routing using longest prefix of hierarchical names is used in CCN. On the other hand, a rendezvous function is used, within a specified scope to solve the issue in PIRSP. For transport and caching, CCN transport using named based routing; finds cached object through local search as well as on the path to the publisher. For PIRSP, transport routing and forwarding use separate forwarding identifier. So, the two projects vary in their design concepts. Similar variations are also found in other projects like DONA [15], NetInf [16] etc.

Though at some points like naming, security, etc deserve early initiative of standardization, the interest of this article on standardization issue is on the major network player interoperability perspective. Now the question is at which point or level of a network (or operator) we expect to communicate with other network (or operator) in ICN architecture. Before explain that, we recall the brief scenario of the extensive cache structure of ICN. It is evident from the ICN literature that to achieve the optimum benefit from the ICN, all major network players have to establish cache in the strategic locations of their network.

Five types of network players, i, e., Internet Service Provider, ISP (Tier 3 ISPs or Tier 2 ISPs), Transit Network (Tier 1 and Tier 2 ISPs), Content Distribution Networks (CDNs) and Large Scale Savvy content provider and publisher can be considered as major network players in ICN. Among them all network players will install cache other that publishers. How cache deployment can bring benefits to the different network players are explained precisely in paper [4]. Now, we know, in ICN architecture, the content will be retrieved from the nearby content router, CR (cache) whether the CR belongs to the same operator, or not. However, the proprietary nature of most content routing designs in use today makes them undesirable for global use and is in conflict with the Internet open standard philosophy.

Since to achieve the optimum benefit from the ICN design, edge router (CR) compatibility and the interconnection between the CR is very important, we believe, an early initiative should be taken by the standardization body to find the standard for CR. On an abstract level, Information centric networking can be compared to a network with CDN server everywhere. So, the lessons from CDNs can rightly explain the necessity of the standardization of the router to router communication in ICN. As we know CDN interconnection can provide benefits to CDN service providers as well as to end users. Since its legacy, CDNs were realized using proprietary technology, open interface support for connecting with other CDNs has been ignored. Even though several CDNs could be interconnected, it is still challenging to achieve interconnection at operational level such as exchange billing information and so on. CDNi-WG is pursuing solutions to these problems.

To sum we can say that an early standardization initiative on economic and technical phenomenon as well as a policy initiative can significantly contribute in the maturation process of the ICN as future internet architecture.

4 Proposed Content Distribution Model for ICN and Related Conceptual Overview

In this section, we propose rational content distribution model which suit with the Information Centric Network. We use this proposed content distribution model for modeling the game among the network players. In current internet architecture, typically client-server model or Global CDN or combination of both Global CDN and client-server model are used by the majority of the content providers or publishers. In case of clientserver model the publisher use their own or leased servers for their content distribution. The key considerations of the content provider's current content distribution system are cost effectiveness, reliability, scalability, flexibility and fast content distribution. Generally, publisher choose client server model because of the cost effectiveness of the model. Nevertheless, publishers choose Global CDN in the case where large traffic amounts are expected. Besides, in case of live streaming publishers choose Global CDN. We assume that for the ICN architecture, publisher will consider the same parameters (cost efficiency, reliability, scalability etc.) for selecting a contentment distribution model. Like current trend, if the publisher gains more users, for the scalability problem publisher will be interested to buy services from Global CDNs. Besides, in ICN, local ISPs will be equipped with in network cache; hence the response time (user perceived time between sending a request and receiving the response) will be minimized resulting Quality of Experience (QoE) for the users. So, compared to the current internet architecture where cost effectiveness is given the highest priority in choosing a content distribution model by the publisher, in ICN faster content distribution will be equally important.

We propose two possible options for publisher to disseminate its content as follows: In the first notion we assume that the publisher can ignore Global CDN providers and have direct connections with Telco CDN. Telco CDN will ensure the big economy of scale for the publisher by Telco CDN interconnection/Federation based on 'Smart Pipe Concept' [17]. In the second notion we assume that publisher will choose Global CDN for its content disseminations and for big economy of scale. And Telco CDN, Telcos and ISPs will ensure last mile connectivity and optimized delivery of the content where they will receive revenue from Global CDN for providing Quality of Experience (QoE). The scenario is briefly depicted in Figure 1.

Now, in ICN the ISP (or Telco) will serve the content to the client possibly in following ways. Each time after receiving a request from the client, the content is fetched from the source server or the Telco's CDN

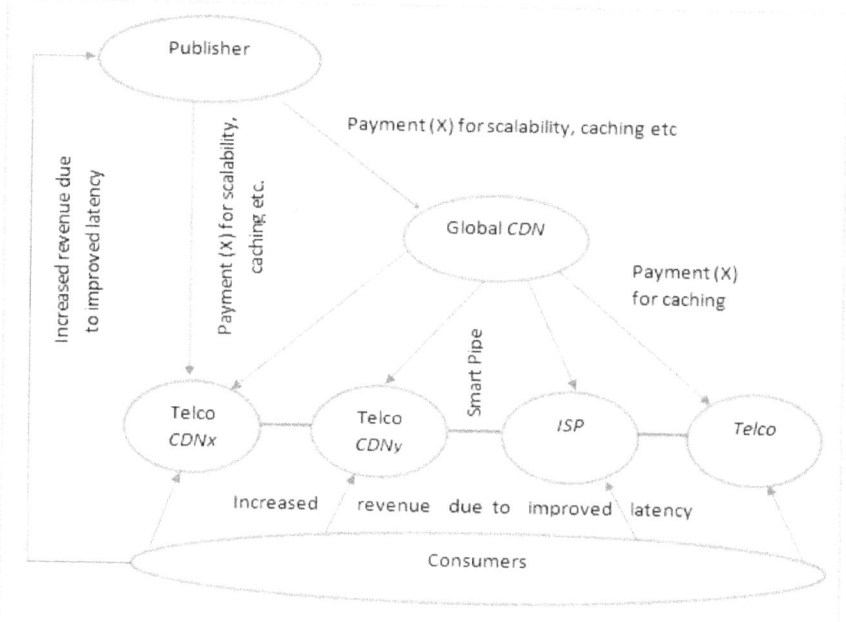

Figure 1 Possible content distribution scenario in ICN architecture

server (its own or other's) or Global CDN's server and serves the client. The other option is to keep the copy (either on-path-caching or off-path caching) of the content in the Content router (of ISP's local network) and serve the client with this cached content for all subsequent requests. Now, typically publisher's content will be managed, cached and hence served by Telcos or ISPs with an objective to minimize the total transit traffic of the Telcos or ISPs. The conventional caching algorithm is that, for the incoming traffic gradually replaces the least popular contents by the contents that are becoming popular. The content popularity typically measured based on its request frequency. Though this strategy is equally valid for ICN, however in addition to that, even though the content is not popular enough, the Telcos or ISPs can provide caching service for the content of selective publishers (due to agreement or good business relations through CDN) by deriving propriety algorithms. The examples of other type of contents (information) like warning of disasters; vaccination information etc. deserve priority treatment (providing on-path/off-path caching service) by the Telcos & ISPs as an obligation of social responsibility.

Figure 1, shows the relationship between different network players in a content distribution model of ICN where publisher choose Global CDN or Telco CDN for its content distribution. Though, the Telcos/ISPs delivery the content to the end user, the publisher cannot pay directly to Telcos or ISPs. It will increase the transaction cost of publisher since in that case publisher has to make a business relation with several/many Telcos and ISPs. So, a rational assumption is that publisher will pay to the Global CDN or Telco CDN in order to publish and delivery of its content and caching. By caching more popular content the Telcos and ISPs can maximize its cash inflow from the customer side. On the other hand, by caching the CDN's (Global or Telco) replicated content Telcos and ISPs can maximize its cash inflow from the CDN side. It is realistic that, content with the similar popularity Telcos and ISPs will be interested to cache a content which can minimize more transit traffic i.e., if the content need to be fetched from the remote server, it will be cached rather the content that will be served by the CDN cache server at the proximity. With this strategy Telcos and ISPs can avoid higher delivery time for content as well. In the end, Telcos and ISPs will face the optimization problem to maximize the revenue to support its client by providing caching service for the popular content and to providing caching service to the chosen publisher's content which is not popular yet but publisher paid for the caching service through Global CDN or Telco CDN.

Now for the ISPs, if they are not intent to institute the caching structure, they must invite the third party to establish cache structure inside their network. Without caching services they cannot provide enhance user experience to their clients hence cannot keep pace with the services of other service providers. Since CDN (Global or Telco) are already doing the alike and they have expertise, they can establish cache functionalities for the small ISPs and maximize their cash inflow.

5 Two Player Game Model for ICN

In this section, we formulate a game theoretic model for the network players for ICN architecture. Firstly we propose a game theoretical model for the two network players; Publisher and Telco CDN. Secondly, we briefly discuss the revenue sharing scenario between the Telco CDN (who has agreement with the publisher for content dissemination) and other peering Telco CDNs (*or Telcos or ISPs*).

The game is based on the proposed content distribution model that we described in Section 3. The whole game is actually motivated by the effort of each player with an intention to maximize its revenue. We assume there is no constraint on the agreement that may accomplish among players. And each network player is considered as an optimizing agent and it expects a reaction from other players to its own action, thus its payoff is determined by other player's actions as well. Since game theory cannot tell us what payoff does each outcome yields, we assign rationale payoff for each player for each configuration which is defined in course of modeling the game. The payoffs are measured in the same units.

5.1 Game Model between Publisher and Telco CDN

Since we assume there is no constraint on the agreement that may accomplish among players, both the network players can choose their position freely. If Publisher serves the content with priority then Telco CDN can serve it with priority or without priority. Similarly, if Publisher treats contents without priority, Telco CDN can be agree to serve it with priority or it can deny serving with priority. Definition of priority for Publisher and Telco CDNs are as follows. In Publisher's point of view priority means increasing load. In Telco CDN's point of view, providing priority service means, choosing the best underlying internet paths that one caching server of Telco CDN has to go through to communicate with another Telco CDN's (or Telcos or ISPs) caching server or the customer website and avoid the congested routes. Moreover to ensure priority service (i.e., QoE) Telco CDN will cache the content in its local content routers (within the Telco CDN's network). Besides, it will pursue (by revenue sharing) other peering Telcos or ISPs to do the same.

The Figure 2 depicts the choices and corresponding payoffs for the Publisher and Telco CDN. If Publisher serves the content without priority (normal load), and Telco CDN also treat it as non-priority then both Publisher and Telco CDN have zero payoffs. Caching contents within the Telco CDN (and its peering Telcos & ISPs) will be on the basis of its popularity index and other conventional strategy. When Publisher serves content without priority (normal load) but Telco CDN treats it with priority, Publisher will have a zero payoff as before. On the other hand Telco CDN will get a negative payoff of -1. This is an opportunity cost for Telco CDN. It is because Telco CDN could waive the priority (i.e., omit the caching service, stop paying the peering Telcos or ISPs etc.) and use the resource (cache/memory space, monetary transfer to peering Telcos/ISPs etc) for other Publisher's contents. Now, if the Publisher serves the content with priority (increased load) and Telco CDN also treat the content with priority then the Telco CDN will receive a higher payoff

		Strategy of Publisher	
		Priority	*Non-Priority*
Strategy of Telco CDN	*Priority*	(1 , 2)	(-1 , 0)
	Non-Priority	(0 , -1)	(0 , 0)

Figure 2   Payoff vectors for Publisher, CDN and TELCO CDN without QoE guarantee

of 2. The reason behind the higher payoff of Telco CDN can be explained in this way. Since the publisher can earn more revenue due to the improved latency of accessing its content, publisher will be happy with the Telco CDN's service hence would possibly renew the contract. The Telco CDN's payoff will be 1 in this case. On the other hand, if the Publisher serves its content with priority (increased load) but Telco CDN treats it as without priority; publisher will lose revenue and (possible lose clients as well; which results a deterioration of the business relationship between the publisher and the Telco CDN. So, in this case the Telco CDN would have a negative payoff of –1.

5.2 Revenue Sharing between Telco CDN and Peering Telco CDN (or Telcos or ISPs)

In this section, we clarify the scenario of revenue sharing between the host Telco CDN (who has agreement with the publisher for content dissemination) and its peer (other Telco CDNs or Telcos or ISPs). The host Telco CDN will earn money from the Publisher and rely on other ISPs or Telcos as a means for last mile connectivity. The peering parties (Telco CDN, Telcos or ISPs) will install extensive cache (Content router) to serve the content to the clients with improved latency hence will provide the quality of experience. Now only charging the end user is not enough for the Telcos (and ISPs) to install and maintain the sufficient caching infrastructure to provide quality of experience. As we know the Telco 1.0 business model of charging the end-user for per minute or for per Megabyte is under pressure and new business models for the distribution of content and transportation of data are being developed and appreciated. Operators will need to be capable of charging different network players including end-users, service providers, third-parties (example advertisers) on a real-time basis for provision of guaranteed quality of service (QoS) [18]. So, according to our proposal the peering parties (Telco CDN, Telcos or ISPs) will charge the host Telco CDN for providing the caching service for the content that it access from that host Telco CDN. And the Smart pipes will be used to Charge appropriately for use of the network [17, 18].

6 Three Player Game Model for ICN

In this section, we provide two separate game models for the three network players; Publisher, Global CDN and Telcos (or ISPs). This game model is also based on the content distribution assumptions described in Section 3 and it holds the other necessary assumptions mentioned in Section 4 as well. In Subsection 6.1, we present the game where there is no agreement for providing Quality of Experience (QoE) among the network players of ICN. We termed this game as a Basic game model among the players. In Subsection 6.2, we consider the Quality of Experience (QoE) guarantee is an integrated part of the agreement among the network players and accordingly we formulate the game model. In Subsection 6.3, we analyze the business relationship among the network players in case of Quality of Experience provision.

6.1 Basic Game Model between Publisher, Global CDN and Telcos (or ISPs)

The Figure 3 shows the choices and corresponding payoffs for the publisher, Global CDN and Telcos (or ISPs). All the network players can choose their position freely. Publisher can serve its content with priority requirements (increased load) or non-priority (with normal load). The case of increasing load by the publisher can be defined as priority for the publisher's point of view. And when the publisher uses the normal load it is termed as non-priority from publisher's perspective as we defined in Section 5.

If publisher increase the load (priority), Global CDN can treat its content with priority or Global CDN can deny giving priority treatment to the its content. Similarly, if Global CDN serves the content with priority, Telcos or ISPs can treat it with priority or without priority. When publisher's content is

		Strategy of Publisher	
		Priority	*Non-Priority*
Strategy of Global CDN and Telcos (or ISPs)	*(Priority, Priority)*	*(1, 1 ,4)*	*(-1, -1, 0)*
	(Priority, Non-Priority)	*(0 ,0, -1)*	*(-1, 0, 0)*
	(Non-Priority, Priority)	*(0 ,0, -1)*	*(0, -1, 0)*
	(Non-Priority, Non-Priority)	*(0 ,0, -2)*	*(0, 0, 0)*

Figure 3   Payoff vectors for Publisher, Global CDN and Telcos or ISPs without QoE guarantee

with normal load (without priority); if Global CDN and Telcos (or ISPs) also treat it as non-priority then all three network players will have zero payoffs. If the Global CDN serve the content with priority and the content is assigned as normal or non-priority by the publisher; Global CDN will get a negative payoff of -1. This is an opportunity cost for Global CDN. It is because Global CDN could waive the priority and use for other content of the same publisher or other publisher which deserves priority service. If Telcos (or ISPs) serves the same content with priority then it will also have a negative payoff of -1, for the similar reason that Global CDN earns a negative payoff. If the publisher increases the load (the case of priority) and if Global CDN and Telcos (or ISPs) also serve the content with priority then the publisher will receive a high payoff of 4. The publisher can earn more revenue, and would possibly renew the agreement with the Global CDN; similarly Global CDN will be happy with the Telcos (or ISPs). So, the Global CDN and Telcos (or ISPs) will achieve a payoff of 1. On the other hand, if the publisher marks its content as priority but Global CDN and Telcos (or ISPs) do not give priority service to the content; the client will not be happy with the service and publisher will end up with loosing revenue and loosing clients. So, in this case the publisher would have negative payoff.

6.2 Game Model between Publisher, Global CDN and Telcos (or ISPs) with Quality of Experience (QoE) Choice

If we take a short look at the kinds of contract that prevail between Global CDN and publisher; we find that from its legacy Global CDN hardly guarantee the publisher about how long it will take for a client to download contents or access contents. Usually, similar to many other internet services, Global CDNs used to sign the fixed fee agreements with no precise service guarantee. However, it changes over time and the contracts include service guarantee like percentage of "up time" or percentage of request that must be served within a specific time stamp. In case of ICN architecture where the service time of a request will be noticeably reduced, it can be easily foreseen that the Quality of Experience guarantee will be an integrated part of the contract. Note that, Global CDN has control over its caching servers and the way they are connected among each other and connected to the business sites. However, Global CDN does not have any control over the caching strategies (on-path or off-path) of the Telcos and ISPs.

We use the term Quality of Experience (QoE) in case of the Telcos and ISPs and Quality of Service (QoS) in case of Global CDN. Since the Quality of Experience (QoE) is used for evaluating the user experience and it is a measure of the customer satisfaction level, we relate it with the deal of Telcos and ISPs. On the other hand since Quality of Service (QoS) cannot evaluate the quality as seen by the end user and its performance indicator is more network-centric rather than user centric we relate QoS with the deal of Global CDN. Though the term Quality of Experience deserves more holistic evaluation; for the simplicity in the context of our proposed game theory, we define it as follows. When Global CDN give priority to particular publisher's content, it means it is providing the Quality of Service (QoS) for the content. And when Local Telcos and ISPs sufficiently caches the content to realize the optimized delivery of contents to fulfill the user satisfaction, it is providing the QoE for its client for that particular publisher's content which it receive from Global CDN.

Now let us see, how the choice of Quality of Experience (QoE) guarantee works for the three network players. Global CDN will give a guarantee for attaining quality of experience to the clients of the publisher's content, if publisher pay the fee (negotiated) to Global CDN. Global CDN will do the agreement with the publisher on behalf of Telcos and ISPs as well. It is because; publisher will try to minimize the transaction costs to set up business relation with Telcos and ISPs. The cost of publisher would be 2 and adds directly to

Global CDN's payoffs. And similarly, Global CDN makes QoE agreement with Telcos and ISPs which cost Global CDN 1 and adds directly to Telcos and ISP's payoff. Figure 4 shows the new payoff.

The agreement of the QoE normally works as follows. If Global CDN meets the QoS (and also Telcos/ISPs meet the QoE) then Global CDN will receive an additional payoff of 2 as bonus. The publisher serves its customers well, makes an additional profit of 4 because of it. Its net profit hence will be 2 after paying the bonus to Global CDN. If the QoS doesn't meet then the Global CDN will have to refund the QoE fee plus pay a penalty of 1 to publisher. And the Telcos (or ISPs) also get a bonus of 1 from Global CDN for meeting the QoE and refund the agreement fee plus penalty of 1 in case of failure to meet QoE. Figure 5 depicts the detailed scenario.

6.3 Analysis of the QoE Choice for Publisher, Global CDN and Telco CDN

First of all, we are interested to see if the game with QoE choice reaches a stable situation, where none of the network player is interested to change its strategy provided that all other players keep their strategy unchanged. This situation is called Nash Equilibrium. Note that the game can have more than one Nash Equilibriums. By analyzing Figure 5, we see that, by choosing

		Strategy of Publisher	
		Priority	*Non-Priority*
Strategy of Global CDN and Telcos (or ISPs)	*(Priority, Priority)*	*(2, 2 , 2)*	*(0, 0, -2)*
	(Priority, Non-Priority)	*(1 ,1, -3)*	*(0, 1, -2)*
	(Non-Priority, Priority)	*(1 ,1, -3)*	*(1, 0, -2)*
	(Non-Priority, Non-Priority)	*(1 ,1, -4)*	*(1, 1, -2)*

Figure 4 Payoff vectors for Publisher, Global CDN and Telscos or ISPs for making QoE guarantee agreement

		Strategy of Publisher	
		Priority	*Non-Priority*
Strategy of Global CDN and Telcos (or ISPs)	*(Priority, Priority)*	*(3, 3 , 4)*	*(0, 0, -2)*
	(Priority, Non-Priority)	*(0 ,-1, 0)*	*(0, 1, -2)*
	(Non-Priority, Priority)	*(-2 ,1, 0)*	*(1, 0, -2)*
	(Non-Priority, Non-Priority)	*(0 ,-1, -1)*	*(1, 1, -2)*

Figure 5   Payoff vectors for Publisher, Global CDN and Tecos or ISPs for guaranteed QoE

the priority option the publisher's payoff becomes (4, 0, 0, −1) and without giving priority payoff vector is (−2, −2, −2, −2). Similarly, in case of Global CDN, the payoff vector for priority choice is (3, 0, 0, 0) and payoff vector without priority is (−2, 0, 1, 1,). And Telcos or ISP's payoff vector by choosing priority is (3, 1, 0, 0) and payoff vector which it can earn by denying priority (−1, −1, 1, 1). Now if we consider a particular configuration for example all player choose priority, then the payoff for Publisher, Global CDN and Telcos (or ISPs) becomes (4, 3, 3) which is the highest payoff for all. So, in this configuration none of the players will change its strategy. Thus we conclude that there will be at least one Nash Equilibrium in the game that we modeled for the quality of service option.

Now, we explain how the QoE choice is the win-win outcomes for all the three network players. We find the solution by using forward deduction scheme. And the comparison will be between the matrices of Figure 3 and Figure 5. First we can remove the column titled non-priority for publisher from the Figure 5. Because, it could have stay with no quality of experience and receive a payoff of 0 instead of −2. Now publisher will choose priority (payoff 4), rather than choosing general (payoff 0). Now publisher is left with two payoff vector. One is go for quality of service guarantee and earn the payoff of $(4, 0, 0, -1)$ and other is go for $(4, -1, -1, -2)$. Obviously, publisher will go for the first one. Because, it will give it more insured service. Similarly, the quality of service agreement will results Global CDN and Telcos (or ISPs) with a payoff 3 which is higher than the payoff 1 which they receive without going to QoE agreement. Thus the QoE agreement is a win-win situation for all three network player.

7 Live Streaming Media Broadcast in ICN Perspective

Achieving high quality and high volume web casts is very challenging. The amount of simultaneous request over such webcast is typically extremely high (for example the expected number of viewers for Tokyo 2020 Olympic). Currently, CDN uses multicasting which is a proven solution. In ICN structure this multicasting can be realized more efficiently. However, the interest of this article is on the economic part of the multicasting for ICN. In multicasting, many request come from the same paths. So, it is more economical to serve the requester by sending the streaming files one time along the path and get as many users as possible to receive the file. Within the extensive cache structure of ICN, the users expressing the interest for the same streaming files can be grouped and sub-grouped optimally which lead to find Steiner minimal multicast tree resulting optimum bandwidth consumption. Hence, by exploiting interest aggregation mechanism and extensive cache structure, there is a great potential for ICN architecture for offering efficient live streaming media broadcasting service. Now, we will show how the CDN and ISP can make their pricing decision about broadcasting the live streaming media over the internet. We show contribution margin for each client subscribing the live stream web cast in the ICN architecture.

First, we consider the lowest three paths of the Figure 6.

For Cache x_1, combined client value of cache x_1,

$V(\text{Cache } x_1) = (x_1 + x_2 + x_3)$ and the cost $C(\text{Cache } x_1)$ is

n_1. If $V(\text{Cache } x_1) > C(\text{Cache } x_1)$ then the path from

Cache x to Cache x_1 will be accepted. Similarly, the Cache x has the combined client value of $V(\text{Cache } x)$ will be

$$\{(x_1 + x_2 + x_3) + (x_1 + x_2) + (x_1 + x_2 + \cdots + x_n)\}$$

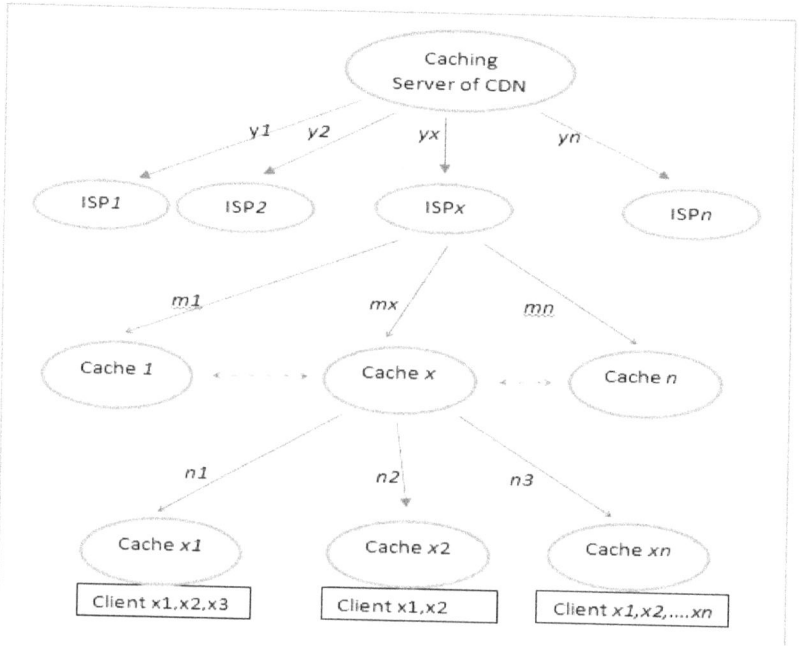

Figure 6 Multicasting decision tree for the real time media streaming for CDNs and ISPs

And the cost is C (Cache x) = $(n_1 + n_2 + n_3)$

Now, pie created at cache x can be calculated as value–cost i.e., Pie (Cache x) = V (Cache x) – C (Cache x). Now, the cost of path towards Cache x is m_2. So, the path, m_2 will be accepted by ISPx if $m_2 <$ Pie (Cache x). Thus, ISPx can calculate its accumulated pie value to make decision whether it is feasible to make contract with CDN. Similarly, CDN will make agree to cast the live stream by comparing the cost and pie value it can accumulate from its different path towards ISPs. We can calculate the contribution margin for each client as well.

Marginal contribution (Client i) = Pie (Root(Client i)) – Pie (Root (Client i)\Client i).

For example, let us calculate marginal contribution of Client x_1 (connected to Cache x_1). First if Client x_1 is not present, we have to see whether the path n_1 will be accepted by the Cache x or not. If $x_2 + x_3 < n_1$ then the path will be rejected i,e, not served. In that case, accumulated pie will be

$$\{(x_1 + x_2) + (x_1 + x_2 + \cdots + x_n) - (n_2 + n_3)\}$$

Hence, the marginal contribution of client, x_1 is

$$\{(x_1 + x_2 + x_3) + (x_1 + x_2) + (x_1 + x_2 + \cdots + x_n) - (n_1 + n_2 + n_3)\} - \{(x_1 + x_2) +$$
$$(x_1 + x_2 + \cdots + x_n) - (n_2 + n_3)\}$$
$$= \{(x_1 + x_2 + x_3) - n_1)\}$$

With this simple mathematical analysis we have shown that the network player like CDNs and the ISPs can find the fair pricing for Live Streaming Media broadcast in ICN.

8 Conclusion and Future Work

Information centric network will offer a novel, enhanced and enrich user's experience for accessing content due to improved latency, which results with increased publisher's revenue. With game theoretical model, we have shown how the major network players like Global CDN, Telco CDN, Telcos and ISPs can have a fair share of this increased publisher's revenues. Moreover we address the issue of the guarantee of the Quality of Experience (QoE). With an analysis we have shown that our proposed game reach to Nash Equilibriums and the agreement for Quality of Experience (QoE) choice is a win-win situation for all network players; Publisher, Global CDN and Telco CDN, Telcos and ISPs.

Acknowledgment

This research was supported by a grant-in-aid from the High-Tech Research Center Project of the Ministry of Education, Culture, Sports, Science and Technology (MEXT), Japan.

References

[1] D. Trossen, and G.K. Alexandros. "Techno-economic aspects of information-centric networking." *Journal of Information Policy* 2 (2012).

[2] G. Xylomenos, X. Vasilakos, C. Tsilopoulos, V. A Siris, and G. C. Polyzos, (2012). Caching and mobility support in a publish-subscribe internet architecture. *Communications Magazine, IEEE*, *50*(7), 52–58.

[3] V. Jacobson, M. Mosko, D. Smetters, and J. J. Garcia-Luna-Aceves, "Content-centric networking: Whitepaper describing future assurable global networks." Response to DARPA RFI SN07-12, 2007.

[4] D. Trossen (ed.), "Architecture definition, component descriptions, and requirements." PSIRP project, 2009.

[5] D. Trossen, M. Sarela, and K. Sollins, "Arguments for an Information Centric Internetworking Architecture." ACM SIGCOMM Comp. Commun. Review, vol 40. No. 2, Apr 2010 pp 27–33.

[6] P. Kwadwo Agyapong, M. Sirbu, "Economic Incentives in Information-Centric Networking: Implications for protocol Design and Public policy." IEEE Communication Magazine December, 2012.

[7] J. Rajahalme, M. Sarela, P. Nikander, S. Tarkoma, "Incentive-Compatible caching and Peering in Data-Oriented Networks", Proc. 2008 ACM CoNEXT Conf. Dec 2008, pp. 62:1–62:-6.

[8] D. Niyato, A.V. Vasilakos, K. Zhu, "Resource and Revenue Sharing with Coalition Formation of Cloud Providers: Game Theoretic Approach." IEEE/ACM International Symposium on Cluster, Cloud and Grid Computing, 2011.

[9] Ng, Chaki, "Game Theory Applications for Content Delivery Networks."

[10] Marco Nicosia, "Internet Video: New Revenue Opportunity for Telecommunications and Cable Providers", Cisco Internet Business Solutions Group (IBSG), July 2010.

[11] Karl, Holger, and Benoit Tremblay. "Document Properties: Document Number: FP7-ICT-2009-5-257448-SAIL/D2. 8." (2012).

[12] B. Chun, K. Chaudhuri, H. Wee, M. Barreno, C. Papadimitriou, and J. Kubiatowicz, "Selfish Caching in Distributed Systems: A Game Theoritic Analysis," PODC, 2004.

[13] P. Golle, K. Leyton-Brown, and I. Mironov, "Incentives for sharing in peer to peer network.", In ACM Conference on Electronic Commerce, 2001.

[14] Arifuzzaman, M.; Keeping Yu, Sato, Takuro, "Content Distribution in Information Centric Network: Economic Incentive Analysis in Game Theoretic Approach", ITU Kaleidoscope conference, Saint Petersburg, Russian Federation, 3–5 June 2014.

[15] T. Koponen, M. Chawla, B. G. Chun, A. Ermolinskiy, K. H. Kim, S. Shenker, I. Stocia, "A Data-oriented (and Beyond) Network Architecture." *Proc. SIGCOMM* '07, Kyoto, Japan, Aug 27–31, 2007.

[16] B. Ahlgren *et al*, "Second NetInf Architecture Description." FP7-ICT-2007-1-216041-4WARD/D6.2.

[17] Youngseok Lee, "Telco CDN interconnection for Global Content Exchange", ITU Workshop on "Bridging the Standardization Gap", Vientiane, Lao People's Democratic Republic, 30–31 July 2012.

[18] The value of 'Smart' Pipes to mobile network operators, Chris Barraclough, STL Partners/Telco 2.0, November 2011.

Biographies

M. Arifuzzaman received the B.Sc. degree in Computer Science & Engineering from Bangladesh University of Engineering and Technology (BUET) in 2001. He worked as an Assistant professor at IBAIS University, Dhaka, Bangladesh from 2001 to 2005. After that he joined in the Bangladesh Civil Service in 2006 and worked as an Assistant secretary to the Government of the People's Republic of Bangladesh till 2010. He has completed Masters in Global Information and Telecommunication Studies from Waseda University, Tokyo, Japan in 2012. Now he is a PhD candidate at GITS of Waseda University. He received many awards including the best paper award in the ITU Kaleidoscope Conference, Cape Town, South Africa, 12–14 December 2011. His research interests lie in the area of Communication protocols, wireless ad-hoc and sensor networks, Next Generation Mobile communication systems and Future Internet Architecture. He is a student member of IEEE.

K. Yu was born in China, on January 1988. He received his B.E. and B.Admin. degree from Sichuan Normal University, Sichuan, China in 2010 and University of Electronic Science and Technology of China, Sichuan, China in 2010, respectively. He received his M.Sc. degree in Wireless Communication from Waseda University, Tokyo, Japan in 2012. Currently, he is a Ph.D. candidate at Graduate School of Global Information and Telecommunication Studies (GITS), Waseda University, Tokyo, Japan. He is a student member of IEEE. His research interests include smart grid, content-centric networking and their information security.

T. Sato received the B.E. and Ph.D. degrees in Electronics Engineering from Niigata University in 1973 and 1993 respectively. He joined the Research and Development Laboratories of OKI Electric Industry Co., Ltd., Tokyo, Japan in 1973 and he has been engaged in research on PCM transmission equipment, mobile communications, data transmission technology and digital signal processing technology. He developed wideband CDMA system for personal communications system and joined the PCS standardization committee in USA and Japan. He contributed in high speed cellular modem standardization for ITU, 2.4GHz PCS standardization for ITA and wireless LAN standardization for IEEE 802.11. He was a Senior Research Manager and Research Director in Communication Systems Laboratory of OKI Electric Industry Co., Ltd. He served as a professor of Niigata Institute of Technology from 1995 and he researched on CDMA, OFDM, personal communication systems and related area. In 2004, he joined as a professor of GITS at Waseda University and currently serving as a Dean of the Graduate School of Global Information and Telecommunication Studies (GITS), Waseda University. His current research interests include Wireless Sensor Network, Mobile IP Network, ICT in Smart Grid, 4G mobile communication systems. He is senior member of IEICE and IEEE.

Qoxcloud: A Cloud Platform
for QoE Evaluation

Eduardo Saiz, Eva Ibarrola, Leire Cristobo and Ianire Taboada

Faculty of Engineering of Bilbao. University of the Basque Country
UPV/EHU, Spain
E-mail: {eduardo.saiz, eva.ibarrola, leire.cristobo, ianire.taboada}@ehu.eus

Received: 9 December 2014;
Accepted: 9 March 2015

Abstract

The current financial situation, together with the new market conditions, has led to major changes in the ICT sector over the last few years. Many services that in the past were only offered by operators are now held by third parties through the cloud, which has caused a shift towards new business models at the expense of a more traditional market. Furthermore, these economic changes have conducted to socio-cultural transformations with great impact on the user's behaviors.

In this paper a cloud platform for the measurement and evaluation of the Quality of Experience (QoE) is presented. The platform is based in a model (QoXphere) that ensures the user satisfaction in terms of Quality of Service (QoS) and the provider economic benefits, as demanded in current market situation. In addition, the proposed cloud architecture intends to help in the advance of the work item recently opened in ITU-T to establish an ITU recognition procedure of testing laboratories with competence in ITU-T Recommendations by providing a unified cloud environment in which to validate them.

Keywords: QoS, QoP, QoE, QoBiz, QoX, Cloud.

Journal of ICT, Vol. 2, 223–246.
doi: 10.13052/jicts2245-800X.232

1 Introduction

The continuous economic growth in the telecommunications sector over the last decades has slowed down in recent years. The financial crisis has caused a drastic reduction in economic activities and investments in this area and this situation has promoted new business models. Many of the services that had been only offered by operators over the last years are now being provided by third parties through the cloud, driving the rise of this model as compared to the traditional sort. (Figure 1).

These new models have also led to a socio-cultural transformation with great repercussion on the users of this sector. The user's behaviour has evolved as a result of this globalization progress. Nowadays, full time connectivity and ubiquity are highly required by users, who are becoming more demanding in terms of capacity and quality of service (QoS). These changes are significantly responsible in the boosting of the cloud service model. Nevertheless, the proliferation of the cloud services can lead to the existence of a 'nebula', from both the point of view of the user and the provider, when implementing policies to manage the demanded QoS.

Since user-centric viewpoint in QoS evaluation is becoming of greater importance nowadays, a new approach to what should be addressed for an optimal QoS management is being embraced. This approach implies an increasing complexity in terms of QoS measurement that must be taken into

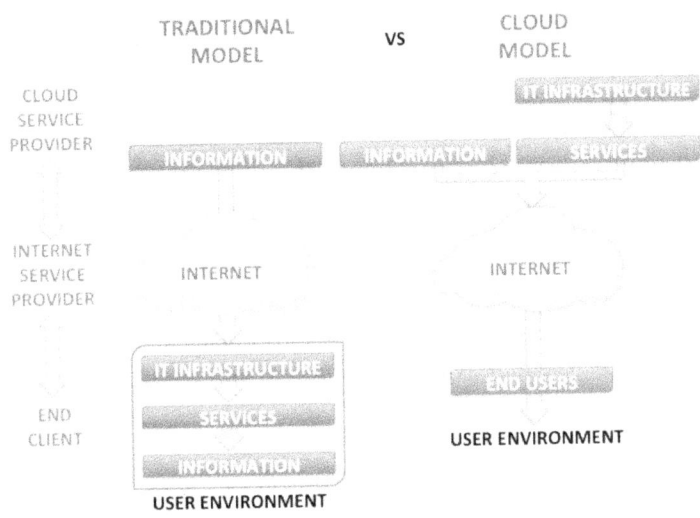

Figure 1   Traditional service model vs. Cloud services model

account. In this regard, new QoS-related terms have emerged in the last years, in both the standardization bodies and the scientific environment. Terms such as the quality experienced by the client (QoE, Quality of Experience) [1], the quality perceived by the user (QoP, Perceived QoS) [2] or the quality of business (QoBiz) [3] have helped to transmit in a more accurate way, the new dimension of what QoS should include. Recently, all these concepts have lead to a new term, which encompasses them all under the name of QoX [4].

In this scenario, the new cloud services model can complicate even more the QoS management, although it can also provide significant benefits in its evaluation. Namely, the cloud itself can be used as a mean for the globalization of the QoS measuring tools and results, which might be of great benefit to the different stakeholders of the sector.

In this paper we present a cloud platform for the measurement and evaluation of the QoS and QoE in the telecommunication services. This platform aims to aid in the new ITU proposal to establish testing laboratories all over the world with the ITU recognition so the measurement platforms and their results can be shared by all the telecommunications community.

The remainder of the paper is organized as follows: Section 2 summarizes related work in both standardization bodies and scientific area. Section 3 describes the new QoXcloud architecture and the QoS model that is adopted as framework [5]. In Section 4, two of the prototypes developed to be included in the cloud QoE platform are presented. Section 5 describes in detail the experiments carried out to validate the prototypes presented in Section 4 and, finally, Sections 6 and 7 contain some conclusions and final remarks.

2 Background

2.1 Standardization Bodies and the Cloud

The cloud services model that is being deployed these days has been defined for the provision and use of any kind of service, application, information and infrastructure with particular emphasis on speed and scalability, according to the needs of each organization, company or user (Figure 2). Given the nature of this model, it's reasonable to see its suitability for organizations with a large number of users, companies or providers. That is the case, for example, of a standardization body. Since this kind of organization has to deal with the provision of multiple services, within its specific coverage ambit, it seems rather appropriate to provide those services through the cloud.

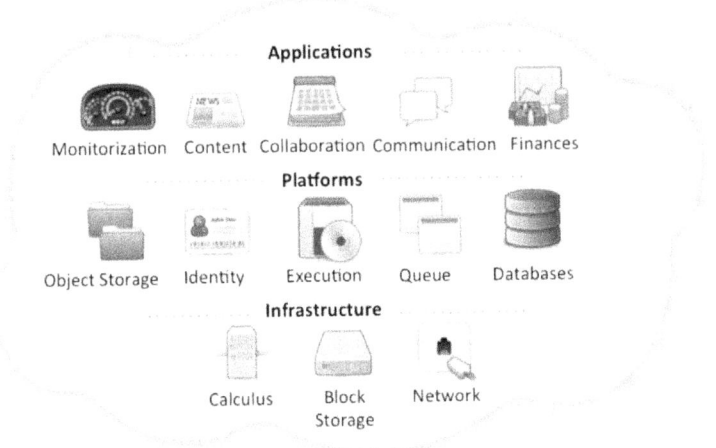

Figure 2   Service provisioning in the cloud

However, and despite the growing activity around this model, there are still very few references coming from the standardization bodies about the management and provision of QoS in the cloud. In particular, the ITU-T SG-13 Study Group, "Future networks including cloud computing, mobile and next-generation networks" [6], has raised some cloud-related questions (Q.17/13, Q.18/13 and Q.19/13) but not with any particular QoS approach. The ITU-T Study Group 11 (SG-11), "Protocols and Specifications" [7, 8], has received proposals showing interest in developing a global architecture (as seen in Figure 3) in which to perform interoperability tests, the validation of recommendations or even the development of measurement procedures for technologies and services that might be of interest for the ITU itself, the Conformity & Interoperability Group [9] or any other collaborating members of the ITU-T study groups such as regulators, operators, providers and clients. This initiative seems to fit very well with the cloud services model. The ITU-T Study Group 12 (SG-12), "Performance, QoS and QoE", has contributed as well to this matter with several proposals for the evaluation of QoS and QoE through the cloud [10, 11]. These proposals will be later discussed since they have been considered in the development of the platform presented in this paper.

The European Telecommunications Standards Institute (ETSI) has also published some reports related to the QoS management in the cloud. Specifically, the TR 103 125 report, "CLOUD: SLAs for Cloud services" [12], deals

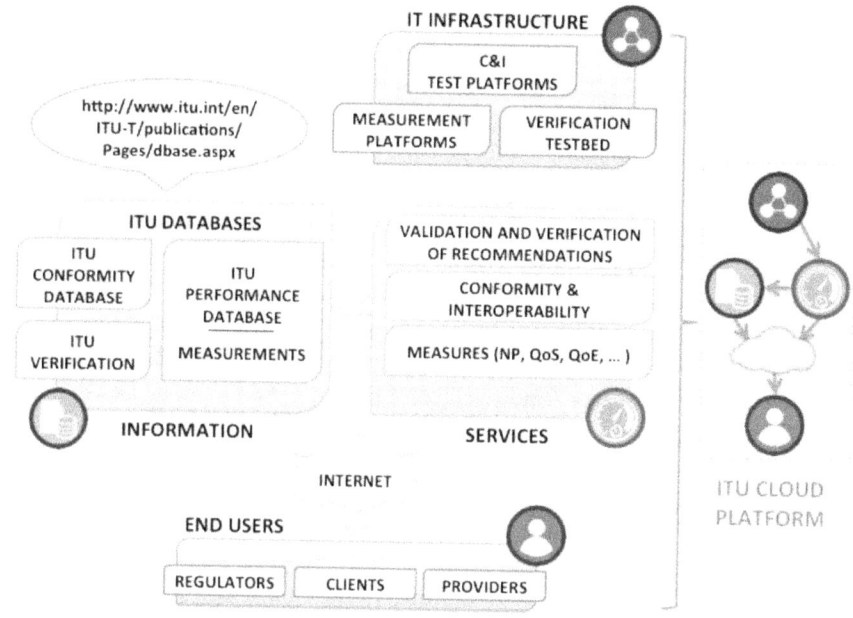

Figure 3 ITU proposed cloud architecture

with the complex problematic of the management of service level agreements in an environment of cloud-based services provisioning.

2.2 Research on QoS Management in Cloud Services

QoS management in the telecommunications environment is a complex issue. Ever since the appearance of newer QoS-related terms that have moved the focus from the objective viewpoint of the Network Performance (NP) parameters to the subjective evaluation of the user's perception (QoP) and experience (QoE), the network resilience (QoR) [13] or the Quality of Business (QoBiz), there is the need of an effective QoS management that does not study each of these terms isolatedly, but as a whole (QoX). The aim is to improve the customer's loyalty and satisfaction, thus ensuring the interests of the different stakeholders (providers, regulators and users).

These factors are also needed in the cloud services model, where resource allocation is quite vital for a proper QoS management. Works by Sharkh [14] in this matter, reveal that QoS provisioning is a rather important challenge. When users execute sensible tasks on a cloud environment, they need a networking service with adequate QoS standards to ensure the successful delivery of their application data. Furthermore, users might require extra guarantees that their information is securely deleted or properly encrypted on the cloud, which implies an extra pressure on the performance and an additional difficulty to comply with QoS requirements.

For this challenge, many authors have made proposals to achieve an effective QoS management in the cloud. According to Cao et al [15], for example, the essence of the cloud services model is to provide network services. As to the user, resources in the cloud should be acquirable anywhere and accessible anytime 'on demand' or by 'pay-per-use'. In combination with Multi-Agent technology and SOA, they propose a cloud architecture that includes physical devices, a cloud services provision layer, cloud services management and a Multi-Agent layer to guarantee QoS in the cloud.

Another interesting work is the one developed by Ferretti [16], in which an architecture specifically designed to respond effectively to the QoS requirements of the cloud applications is proposed (Figure 4). These application requests are handed to a virtual execution environment manager, responsible for optimizing and scheduling the availability of the cloud resources.

Nevertheless, and despite all the recent research, QoS management in the cloud is still a relatively unexplored topic with a great importance that should be taken into account. This, along with the aforementioned ITU proposal of a cloud platform for the validation of recommendations, interoperability tests and other services and technologies, is the primary motivation for the development of QoXcloud, a cloud platform for the measurement and evaluation

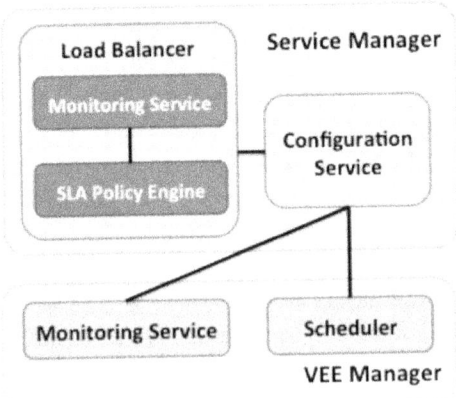

Figure 4   QoS management in the cloud (Ferretti)

of QoS and QoE, that can also help the ITU to establish testing laboratories all over the world, with ITU recognition, for their results and developed platforms to be shared amongst the telecommunications community.

3 QoXcloud Platform

In this section, the QoXcloud platform architecture is presented. Figure 5 shows the three main entities in which the platform is divided.

- *Services Entity*
- *IT Infrastructure*
- *Information Entity*

It must be remarked that the QoXcloud architecture has been designed in order to satisfy the related ITU-T recommendations and standards (Rec. G.1000 [2], Rec. G.1011 [17], Rec. M.3050 [3] as well as Recs. Y.1541, Y. 1542 [18] and Y.1543 [19]).

3.1 Services Entity: QoXphere Framework

QoXcloud services entity is based upon the design of the QoXphere framework [5] (Figure 6), which establishes relationships amongst all the aspects of the QoS (QoX), organized in the following four layers that comply, amongst others, these respective ITU-T recommendations:

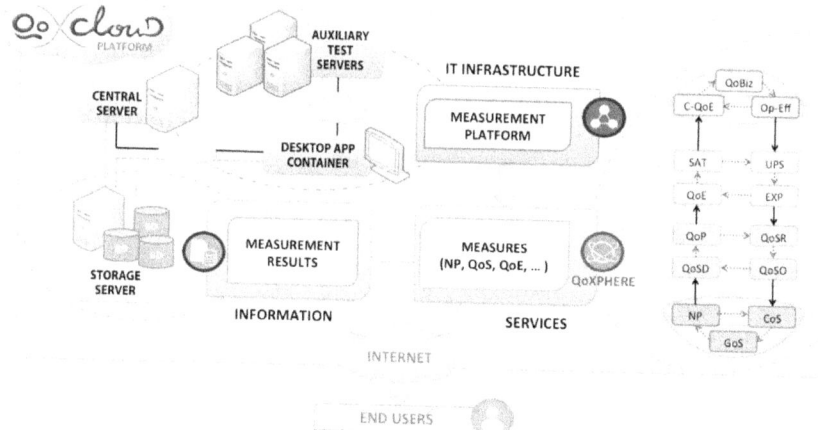

Figure 5 QoXcloud architecture

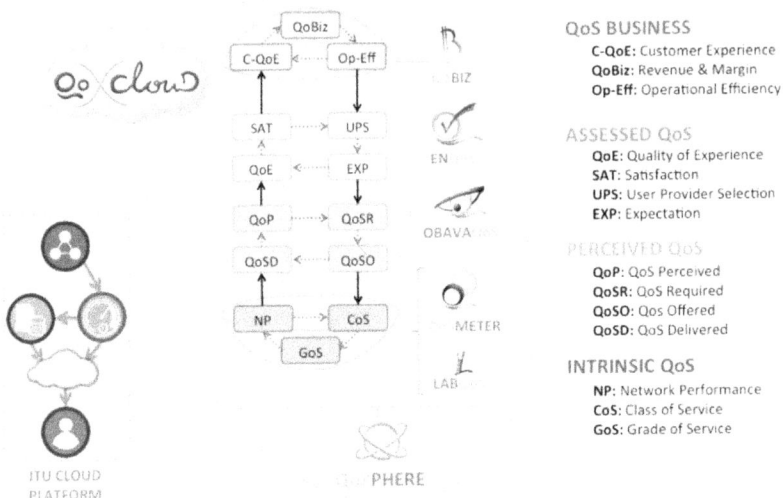

Figure 6   QoXphere framework within the QoXcloud architecture

- **Intrinsic QoS Layer:** This layer analyzes the objective QoS parameters evaluation at the Network Performance level, as defined in ITU-T Y.1540 and Y.1541 Recs. [20, 21].

  For this layer implementation, two platforms has been designed: QoSmeter [22] and LabQoS [23]. The first one is a neutral QoS measurement infrastructure meant to measure objective network parameters through a wide variety of tests that help determine the degree of compliance to the SLA. The second one, LabQoS, is a further development of QoSmeter focused on testing and simulating experimental scenarios.

- **Perceived QoS Layer:** Based on the four viewpoints of ITU-T G.1000 recommendation [2], this layer is considered the 'core' of the QoXphere model, since the whole framework is defined in accordance with it. In fact, QoXphere has been proposed as the reference framework for future updating of the ITU-T G.1000 recommendation [24].

  For this layer implementation, the OBservatory for the Analysis and Validation of QoS (ObavaQoS) has been defined. This subsystem helps determining the required quality indicators (KQI) and performance indicators (KPI) to estimate the QoS perceived in terms of the referred ITU-T G.1000 recommendation.

- **Assessed QoS Layer:** Based on the Xiao's CSAT model [25], this layer evaluates the user's satisfaction based on the user's experience, which is based in its turn on QoP, as evaluated in the previous layer. For this layer's development, a general-purpose on-demand survey management and configuration subsystem is used. This subsystem, named ENQoS, is based upon the ITU-T G.1030 Recommendation [26].

- **QoS Business Layer:** The top layer of the framework is oriented to guarantee the provider's profitability based on the feedback from the other three layers.

  In particular, this layer has been defined to comply the ITU-T E.419 Recommendation [27]. Also, since the Telemanagement Forum has also shown interest in the ITU cloud platform, this layer also accomplishes the references of the TMF GB935 document on business metrics [28].

On a final note to the services entity, it must be said that all the referred subsystems (QoSmeter, LabQoS, ObavaQoS and ENQoS) required being adapted to contemplate their usage in a cloud scenario.

3.2 IT Infrastructure

The IT Infrastructure in the QoXcloud platform is defined as the set of facilities (hardware, software, network devices and the required interconnections) that is capable of running any kind of measurement test within its boundaries. Since different measurement tests require different test scenarios or layouts, the IT Infrastructure has been defined as a modular and scalable distributed architecture, capable of accommodating the widest diversity of tests possible by adding additional test servers or designing different test applications. The aforementioned architecture of QoSmeter [22] (Figure 7) has been used defining the interaction of four distributed entities:

- **Parameter Measurement Services (PMS):** A PMS is the joint of the required test measurement modules and the auxiliary servers towards the tests are held. For example, different endpoint locations for a particular measurement service.

  Any measurement test can be designed for its integration in the QoXcloud IT Infrastructure, by complying with the requirements of the PMS entity and additional authorization policies, as discussed later in the paper.

- **Storage Server (SS):** All results obtained by the PMS modules are persisted in the databases of this server, which is also the main element of the Information entity, as seen in Figure 5.

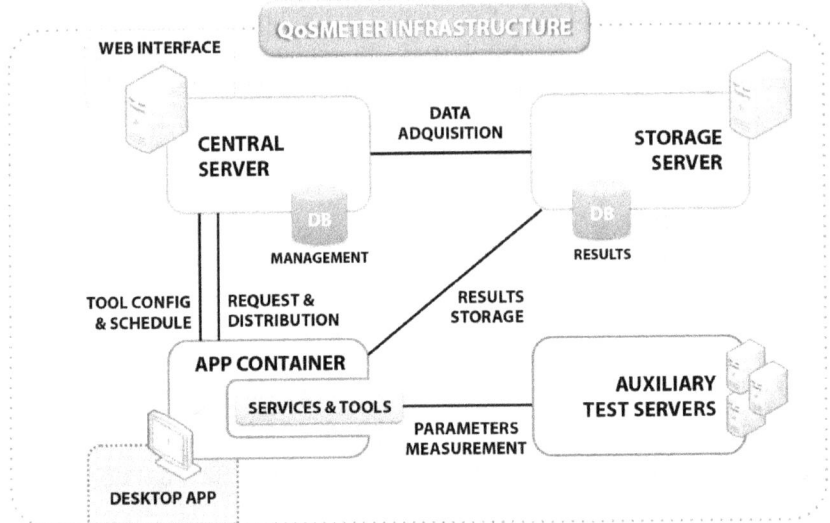

Figure 7   QoXcloud IT Infrastructure: QoSmeter Infrastructure

- **Central Server (CS):** It is the core of the QoSmeter infrastructure and it's in charge of four main actions:
  - *User management:* In charge of the management of user rights according to the roles or groups of belonging.
  - *System management:* Test configurations, reports definition and system status monitoring are managed in this module of the CS.
  - *Scheduling:* This module is in charge of the scheduled execution of tests and their configuration, if applicable.
  - *Report Generation:* Finally, this module is capable of generating a specific report from the required data obtained from the Storage Server.
- **Test App Container (AC):** The fourth entity of the IT infrastructure is an application container from which the measurements are launched (Figure 8). Using a Single Sign On (SSO) procedure, the AC authenticates the user within the infrastructure domains and obtains the list of PMS available for execution.

All the entities interact within a specific authentication and authorization policy. This A & A policy allows the system to distinguish different user types and the specific measurement tests to which they have credentials.

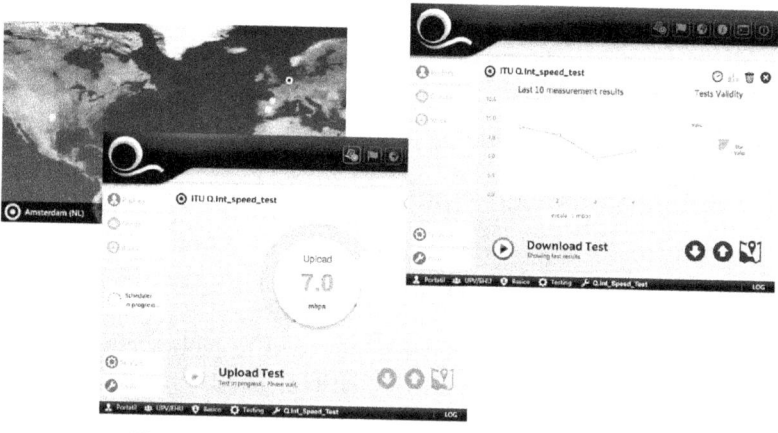

Figure 8 Sample captures of the Test App Container

In this way, a client can execute a particular test related to his contracted service, a provider can use the infrastructure as a neutral reference of its users' results, and a regulator or standardization body can access specific test reports for a particular recommendation to validate.

3.3 Information Entity

As a result of the adoption of the QoSmeter infrastructure as the IT infrastructure, the Information entity of the QoXcloud platform is consequently affected, being the Storage Server of QoSmeter the most adequate implementation for this entity.

The definition of the SS server should be made compliant to the ITU databases in order to get the whole QoXcloud platform validated for its integration.

4 Validation

In this section the two prototypes developed for the validation of the QoX-Cloud platform are described. These prototypes consist of two measurement platforms for the evaluation of the QoE in two of nowadays most used services:

- The Internet access service, as defined in the ETSI EG 202 057-4 guide [29].
- The web service, as defined in the P.STMWeb drafts [10, 11], which were under active development at the time this prototype was proposed.

The definition of the test scenario for these services, as well as the methodological approach to their respective evaluation is explained in the following lines, prior to the presentation of the validation results in Section 5.

4.1 Internet Access Service Prototype

The scope of this prototype (Figure 9) is to measure those parameters that help determine the QoS delivered to users in the Internet access service, as defined in the aforementioned ETSI guide.

As a side note, this same prototype is intended to embrace a new recommendation that is being promoted from the ITU SG-11 in the last couple of months: the ITU-T Q.Int_speed_test draft [30]. Working on the basis of existing regulation and recommendations as the ETSI guide [29], this new recommendation aims to unify the methodology for the measurement of Internet speed by end-users.

Since it is still in an early stage of development, it has been considered appropriate to observe its evolution in the next couple of months in order to adapt the prototype for its requirements. In the meantime, the current

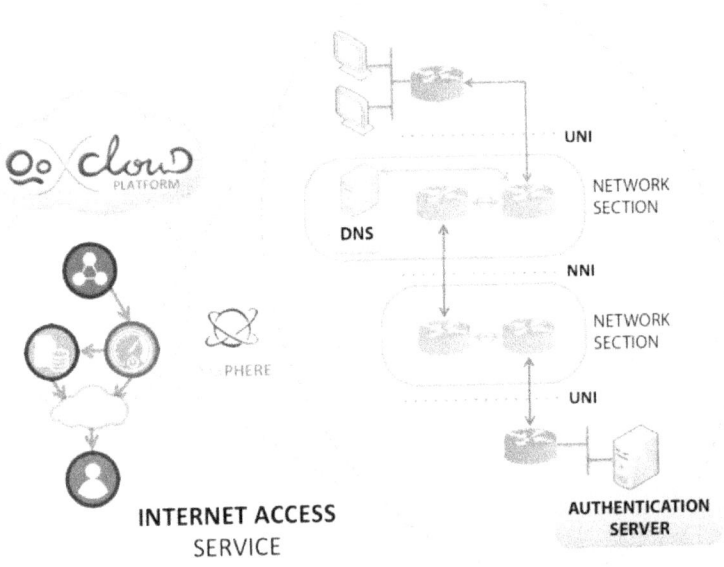

Figure 9   QoS measurement in Internet access service (prototype)

implementation of the prototype has been designed for the measurement of the following parameters:

- **Data transmission speed achieved**: Mean value and standard deviation of the transmission rate achieved (in kbit/s), as well as the maximum and minimum values calculated, respectively, as the highest 95% and the lowest 5% of the data transmission rate achieved (in kbit/s).
- **Delay (one way transmission time)**: Mean value and standard deviation of the delay (in milliseconds).

For these parameters to be measured, a test file has been defined, as specified in Annex D of the guide [29]. That file should consist of incompressible data, so the file remains the same no matter how many distorting/compressing effects the network might introduce.

Also it should be at least twice the size (in kbit) of the theoretically maximum data transmission rate per second (in kbit/s) of the Internet access under consideration. In the particular case of our prototype, that multiplier factor has been set to 4x, to also comply local regulations [31].

To accomplish the measurement of the aforementioned parameters, two subtests have been defined:

- **Speed Test:** This test consists of measuring the elapsed time in transmitting the correspondent test file for both the downlink and uplink transmission rates of the contracted Internet access service, as defined in the previous paragraphs.

  Given the size of the test file for each link, the respective transmission speed achieved is calculated as the size of the test file (in kbit) between the measured time (in seconds) in each link subtest.

  The test scenario consists of the app container installed in the user device and three geographically dispersed servers, located in Madrid (Spain), Dallas (TX, USA) and Amsterdam (The Netherlands). Those auxiliary servers have the required logic to process the download request from the client and response back with the specific file. The container is responsible for measuring the elapsed times and save them in the SS.

- **Latency Test:** This other test consists of the same test scenario presented in the speed test specification, but, in this particular case, the app container launches a poll of 20 ping interactions to each destination host. Times are measured at the beginning and end of each interaction. Please note that, as a soft adoption of the ETSI guidelines, the two-way latency instead of the one-way delay is measured, since time measurements are done on the client side (from the app container).

After the polling is done, the maximum, minimum and average latency values are stored, as well as the jitter (standard deviation), the percentage of losses and an estimation of the Mean Opinion Score (MOS) value based on the E-model, as defined in the ITU-T G.107 Recommendation [32].

4.2 Web Service Prototype

The second prototype designed for the validation of the QoXcloud platform is based on the P.STMWeb Draft [10, 11], which defines a subjective test methodology for the web browsing service. As a complement to this draft, the jointly developed G.QoEWeb Draft [33] enumerates the QoE influence factors in the web browsing service. Both drafts are also intended to update and supplement the ITU-T G.1030 Recommendation [26].

Therefore, the Web Service Prototype takes into account those influence factors, including network bottlenecks and other hardware and software specifications in the test facilities disposed for the validation of the prototype, and adopts the test methodology of the P.STMWeb draft, in accordance to its requirements.

Concretely, the aim of this draft is to define a test to evaluate the QoE of the participants by introducing alterations on the network level, such as in the round trip time (RTT) or in the bandwidth; as well as on the application level (for example, page load times). The draft defines 6 bandwidth conditions (ranging from 64 Kbps to 2048 Kbps) with a fixed RTT of 20 ms. Those conditions are to be applied when navigating two kinds of contents: A news site and a photo gallery site.

The facilities for this test consist of a controlled LAN with a User Device (UD) with a basic web browser, a Network Emulator (EMU) capable of inducing the conditions in the link UD-EMU, a local Content Server (CS) that hosts well constructed news and photo gallery sites, a Page Health Monitor (PHM) to analyze de availability of web contents that might be hosted outside the controlled LAN, and a router, required for LAN interconnection (Figure 10).

The implemented methodology consists of a warm-up phase where users are guided throughout 3 sessions (condition periods of 150 seconds), followed by two rounds of 6 × 2 session iterations per round. This is, each round consists of the 6 bandwidth conditions randomly presented to the user for each of the 2 referred content types. After each session, users are prompted a form in which

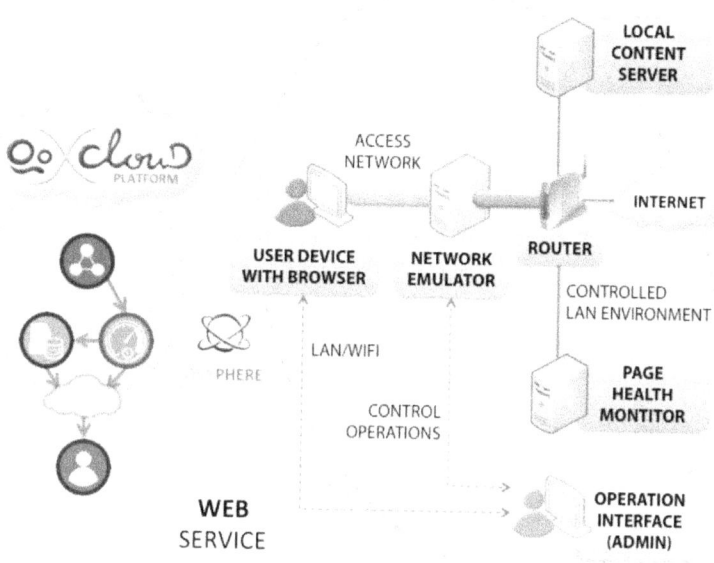

Figure 10 QoE measurement in web service (prototype)

to evaluate their navigational experience to the just finished condition in a MOS scale from very bad (1) to very good (5).

In the particular implementation of our prototype, no external contents to the controlled LAN environment were to be accessed as everything was served from the CS. For that reason, no PHM was used.

5 Results

5.1 Internet Access Service Results

According to the Ministerial Order ITC/912/2006 [31] of the Spanish Government, any Internet Service Provider (ISP) that operates in the Spanish territory and fulfils certain specifications is required to publish results on the levels of QoS they offer and deliver on a quarterly basis.

Therefore, it might be of great interest for the validation of the prototype to present the results of the Internet Access service for a specific ISP obtained from the prototype implementation as compared to the correspondent results published by that same ISP in the Spanish territory, according to the specifications of the Spanish Ministerial Order.

Table 1   Sample comparison of official ISP bandwidth results vs. prototype results

ADSL 6 Mbps	QUARTER		MEAN (Mbps)	PERC 95 (Mbps)	PERC 05 (Mbps)
OFFERED*			4.700	4.900	4.400
	Q4 2012	ISP	5.031	5.077	4.965
		PROTOTYPE	3.970	5.152	2.997
	Q1 2013	ISP	5.022	5.079	4.941
MEASURED		PROTOTYPE	4.008	5.661	2.913
	Q2 2013	ISP	5.002	5.093	4.945
		PROTOTYPE	4.437	4.927	3.608
	Q3 2013	ISP	5.012	5.092	4.944
		PROTOTYPE	4.532	5.118	3.555

* ISP offered values references: Valid since 2010-01

It can be seen in Table 1 that, for the volume of tests performed by users of the QoSmeter platform (integrated in the QoXcloud IT infrastructure) whose Internet access service has been contracted with the ISP to which results are being compared, the obtained results are relatively stable over time but its mean value is below the correspondent results published by the ISP.

This does not mean that the results of the prototype are not valid. Actually, what explains that difference is that users who took part in the Internet access service tests are distributed throughout the Spanish geography and, even for the same contracted bandwidth, they have very different access conditions amongst them, such as different distances to the ISP facilities, different last mile network technologies or even geographical difficulties that might require a high investment by the ISP for a better access provision; while probes that are carried out by the ISP itself are emulated in a controlled environment that ensures the best compliance with the standards required by the Ministerial Order, thus not contemplating real case scenarios.

For this reason, the importance of using a third-party neutral measurement infrastructure can be crucial for a real compliance of current recommendations and regulations. The results obtained in this prototype validation are a good account of it.

5.2 Web Service Results

The Web Service Prototype was tested with a group of 46 participants divided in two groups: experts and non-experts. The non-expert group was composed of 23 undergraduate students of the Telecommunication Engineering Degree at the University of the Basque Country in Spain, whilst the remaining 23 participants were professors and researchers of this university.

The results presented in this section were obtained during the two months the tests were held (October 2013–November 2013). MOS value is calculated within the 95% confidence interval and results are disaggregated by content type and user category.

Figure 11 presents the MOS results for the totality of the participants without group disaggregation. The graphic presents quite linearity for both content categories. However, a lower tolerance to contents with higher page weights can be observed, as photos are rated lower. This effect can be better appreciated in faster conditions, where news text is loaded rather quickly in comparison to the heavy images of the photo contents. When in worse conditions, since loading times are not good even for news contents, both contents' graphics are quite similar.

In Figure 12 the comparison between the results of experts and non-experts is shown. It can be perceived that, in general, experts seem more demanding than non-experts, as well as more accurate when deciding the MOS value. This

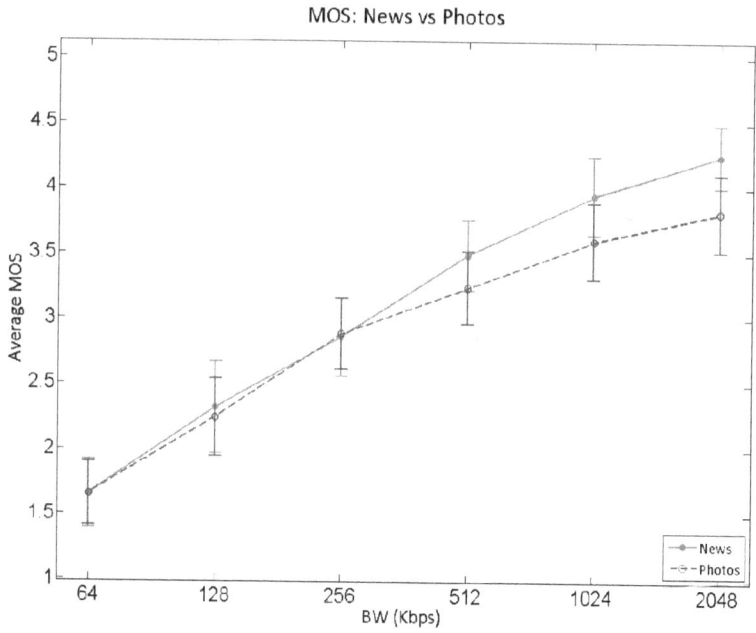

Figure 11 MOS vs. BW (results for 46 participants)

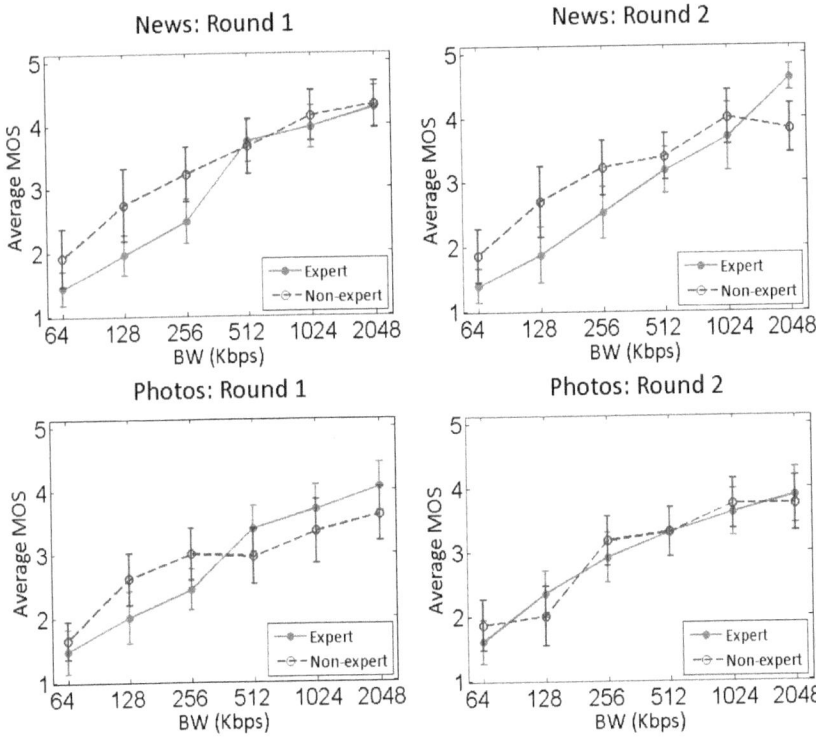

Figure 12   Experts vs. Non-experts MOS results

effect is easily seen in the news category. However, there is no clear tendency for photo contents, as this kind of content is more difficult to evaluate since the final page size depends mainly of the varying sizes of the photos to load.

Another interesting effect observed during the two-month testing period, was derived from the test duration itself: After a few conditions evaluation, the panel of experts seemed to have a very clear decision on what to answer in the subsequent forms within the first seconds of the conditions under way.

For this reason, half of the experts were conducted a short version of the test in which each condition period was reduced from 150 to 100 seconds, in order to compare whether the results showed significant variations or not. Given the similarity of the graphs for both versions of the test, it was concluded that the test durations defined in the P.STMWeb Drafts [10, 11] could be revised downwards for the lesser fatigue of the participants.

Additionally, the fixed range of bandwidth values defined in the draft should be adapted to the existing offer of Internet access speeds in the region or country in which to conduct the tests, in order to obtain more consistent results from the participants with their real web browsing experience.

These proposals, as well as the obtained results and graphics, were included in the contribution to P.STMWeb Draft enhancement, presented for that purpose [34].

6 Conclusions

In this paper a cloud platform (QoXcloud) for the measurement and evaluation of the QoS and QoE in the telecommunication services has been presented.

The proposed platform has been designed on the basis of the QoXphere framework [5], which fully accomplished the most important ITU-T related recommendations. As a consequence of this, the resulting architecture has been designed to fit within the proposal of an ITU global architecture for the performance of interoperability tests and the validation of existing or new ITU-T Recommendations.

The proposed cloud platform has also been conceived with the aim of helping to advance in the work item opened recently in ITU-T to establish an ITU recognition procedure of testing laboratories with competence in ITU-T Recommendations.

The results of the test carried out in both the developed prototypes demonstrate the validity of the platform in terms of a suitable architecture for the evaluation of QoE and the validation of the aforementioned recommendations.

In fact, the contribution to the enhancement of P.STMWeb Draft [34], that was presented following the validation of the Web Service Prototype, was well considered in the final version of the recommendation. The newly published ITU-T P.1501 Recommendation [35] defines lower condition periods and a not fixed but orientative range of bandwidth values for the subjective testing methodology in web browsing, as suggested in the contribution.

Indirectly, the presented contribution also helped in the validation of the G.QoEWeb Draft [33], which resulted with the publication of the new G.1031 Recommendation [36] and the update of the G.1030 Recommendation [26].

7 Future work

As previously mentioned, the ITU-T Study Group 11, with the support of the Conformity & Interoperability Group [37], has promoted in the last couple of months the development of a new recommendation (ITU-T Q.Int_speed_test

Draft [30]), intended to describe a "unified methodology of Internet speed quality measurement usable by end-users". The main objective of this draft is to obtain comparable results from the many Internet testing tools already developed that are accessible from a web browser or desktop application.

Despite being in an early stage, the draft shows some indicators of the suitability of the QoXcloud platform to help serve the ambitious undertaking of the recommendation.

One of those indicators is the test facilities architecture that the Q.Int_speed_test defines. This architecture shares certain similarities with the QoXcloud IT Infrastructure shown in Figure 7, so minor modifications would be needed. As for the methodology itself, the draft defines two measurement tests (an Internet access speed test and an Internet resources access speed test), to which the Internet Access Service prototype presented in this paper could be useful for, if modified adequately to fit the draft requirements.

As a consequence of this, not only studying the evolution of the Q.Int_speed_test Draft in the next couple of months is mandatory, but also a deep analysis of the methodology is required, in order to contribute to the enhancement and validation of the recommendation. Pursuing this objective is one of the main essences of the QoXcloud platform.

8 Acknowledgment

This work has been partially funded by the Spanish Ministerio de Economía y Competitividad (MINECO) under grant TEC2013-46766-R: QoEverage - "QoE-aware optimization mechanisms for next generation networks and services".

References

[1] ITU-T, "P.10/G.100 (2006) Amendment 2 (07/08): New definitions for inclusion in Recommendation ITU-T P.10/G.100", 2008.

[2] ITU-T, "G.1000: Communications quality of service: A framework and definitions", 2001.

[3] ITU-T, "M.3050.1: Enhanced Telecom Operations Map (eTOM) – The business process framework", 2007.

[4] R. Stankiewicz et al, "QoX: What is it really?". Communications Magazine, IEEE, vol. 49, pp. 148–158, 2010.

[5] E. Ibarrola, E. Saiz, J. Xiao, L. Zabala, and L. Cristobo, "QOXPHERE: A new QoS framework for future networks," in ITU Kaleidoscope 2013: Building Sustainable Communities (K-2013), 2013 Proceedings of, 2013, pp. 1–7.

[6] ITU-T, Study Group 13: "Future networks including cloud computing, mobile and next-generation networks", http://www.itu.int/en/ITU-T/studygroups/2013-2016/13/Pages/default.aspx

[7] ITU-T, TD 048 Rev.1 (GEN/11): "Living list of key technologies which are under study in SG11 and are suitable for conformance and/or interoperability testing".

[8] ITU-T, SG-11. COM 11 – C 24 – E: "Proposals for the collaborative QoE/QoS testing platform to advance in "Q.QMS" work item", University of the Basque Country. February 2013.

[9] ITU, ITU Conformity and Interoperability (ITU C&I), from: http://www.itu.int/en/ITU-T/C-I/Pages/default.aspx

[10] ITU-T, SG-12. COM 12 – C 337 – E: "Draft P.STMWeb", A1 Telekom Austria AG. March 2013.

[11] ITU-T, SG-12. COM 12 – C 46 – E: "Draft Test Plan for P.STMWeb". A1 Telekom Austria AG. March 2013.

[12] ETSI, TR 103 125 V1.1.1: "CLOUD; SLAs for Cloud services", 2012.

[13] G. Ghinea et al, "Quality of perception to quality of service mapping using a dynamically reconfigurable communication system", in Global Telecommunications Conference, 1999. GLOBECOM '99, Rio de Janeiro, Brazil, 1999, pp. 2061–2065.

[14] Sharkh, M.A. et al, "Resource allocation in a network-based cloud computer environment: design challenges", Communications Magazine, IEEE, vol. 51, no. 11, pp. 46–52, November 2013.

[15] Cao et al (2009). "A Service-Oriented Qos-Assured and Multi-Agent Cloud Computing Architecture". In M. Jaatun, G. Zhao & C. Rong (Eds.), Cloud Computing (Vol. 5931, pp. 644–649): Springer Berlin Heidelberg.

[16] Ferretti et al., "QoS-Aware Clouds". Paper presented at Cloud Computing (CLOUD), 2010 IEEE 3rd International Conference on, Miami, USA, 5–10 July 2010.

[17] ITU-T. "G.1011: Reference guide to quality of experience assessment methodologies", 2013.

[18] ITU-T. "Y.1542: Framework for achieving end-to-end IP performance objectives", 2010.

[19] ITU-T. "Y.1543: Measurements in IP networks for inter-domain performance assessment", 2007.

[20] ITU-T. "Y.1541: Network performance objectives for IP-based services", 2006.

[21] ITU-T. "Y.1540: Internet protocol data communication service – IP packet transfer and availability performance parameters", 2007.

[22] R. Partearroyo et al, "QoSmeter: Generic quality of service measurement infrastructure," in IFIP Networking 2006, workshop 'Towards the QoS Internet' (To-QoS'2006), Coimbra, Portugal, 2006.

[23] L. Zabala et al, "LabQoS: A platform for network test environments," in ITU-T Kaleidoscope 2011. The fully networked human? – Innovations for future networks and services, Cape Town, South Africa, 2011.

[24] ITU-T, SG-12. COM 12 – C 30 – E: "Proposal on enhancement of G.1000 framework". March 2013.

[25] J. Xiao and R. Boutaba, "Assesing network service profitability: modeling from market science perspective", Networking, IEEE/ACM Transactions on, vol. 15, pp. 1307–1320, 2007.

[26] ITU-T, "G.1030: Estimating end-to-end performance in IP networks for data applications", 2005.

[27] ITU-T, "E.419: Business oriented Key Performance Indicators for management of networks and services", 2006.

[28] TMF GB935, "Business Metrics Concepts and Principles", Release 7.1.1.

[29] ETSI, EG 202 057-4 V1.2.1: "Speech processing, transmission and Quality aspects (STQ); User related QoS parameter definitions and measurements; Part 4", 2008.

[30] ITU-T, SG-11. TD 631-GEN: "Draft Recommendation ITU-T Q.Int_speed_test: Unified methodology of Internet speed quality measurement usable by end-users on the fixed and mobile networks", December 2014.

[31] OM ITC/912/2006: "Por la que se regulan las condiciones relativas a la calidad de servicio en la prestación de los servicios de comunicaciones electrónicas", BOE 31-03-2006, Spanish Gov. 2006.

[32] ITU-T, "G.107: The E model: A computational model for use in transmission planning", 2014.

[33] ITU-T, SG-12. COM 12 – C 34 – E: "G.QoE-Web: Relevant factors and use cases for QoE web". A1 Telekom Austria AG. March 2013.

[34] ITU-T, SG-12. COM 12 – C 0124 – E: "Test results and new proposals for enhancement of Draft P.STMWeb", University of the Basque Country. December 2013.

[35] ITU-T, "P.1501: Subjective testing methodology for web browsing", 2014.

[36] ITU-T, "G.1031: QoE factors in web-browsing", 2014.

[37] ITU, "Measurements of Internet speed", from: http://www.itu.int/en/ITU-T/C-I/Pages/IM/Internet-speed.aspx.

Biographies

E. Saiz received B.S. and M.S. degrees in telecommunications engineering from the University of the Basque Country in December 2009. Since January 2010 he has worked as a research fellow for the Networking Quality and Security (NQaS) Research Group of the University of the Basque Country. In this period he has cooperated in the development of a neutral infrastructure for the deployment of different QoS measurement services. He is currently working toward his Ph.D. degree. His research interests focus mainly on QoS network performance and parameters and QoS/QoE performance measurements.

Dr. E. Ibarrola received the Ph.D. degree in Telecommunications Engineering in 2010 from the University of the Basque Country for her work in the area of user-centric Quality of Service (QoS) management models. She was honored with the Best Thesis Award in Management, Economy and Telecommunications Regulation from the Telecommunication Engineering

Official College in Spain. Prior to joining University of the Basque Country, in January 2000, she worked in the National Network Supervision and Operation Center of Telefónica. She has been participating in different R&D projects and cooperating with different standardization bodies. Her research interests focus mainly on users' oriented (QoS) management models and frameworks.

L. Cristobo received B.S. and M.S. degrees in telecommunications engineering from the University of the Basque Country, Spain, in 2004. She has nine years of communications engineering experience and is currently head of the IT Department at ST3 Elkartea. Since 2007 she has been an associate professor at the University of the Basque Country, where she is currently a Ph.D. candidate. She conducts research in the areas of QoE and quality of business (QoBiz) management processes.

Dr. I. Taboada is a member of NQaS – Networking Quality and Security Research Group from University of the Basque Country (UPV/EHU). She received her degree and her Ph.D in Telecommunications Engineering from UPV/EHU, in 2008 and 2013 respectively. After being working in themes related to PQoS/QoE/QoS assessment in converged networks, for the last three years, her research interests mainly include dynamic scheduling of network resources, especially for QoE optimization in wireless networks. She is familiar with Markov Decision Processes, Gittins index approach and Marginal Productivity indices (Whittle index theory).

Towards the Standardization of Stereoscopic Video Quality Assessment: An Application for Objective Algorithms

José Vinícius de Miranda Cardoso, Carlos Danilo Miranda Regis and Marcelo Sampaio de Alencar

Federal University of Campina Grande (UFCG)
Federal Institute of Education, Science, and Technology of Paraíba (IFPB)
Institute for Advanced Studies in Communications (Iecom)

Received: 11 November 2014;
Accepted: 9 March 2015

Abstract

This article describes a Qt-based C++ application for full-reference stereoscopic video quality assessment, which is platform independent and provides a friendly graphical user interface. The stereoscopic video signals used in the application are based on a two-view model, such as the H.264/AVC standard in the Multiview Video Coding (MVC) profile. In addition, several spatial resolutions are available. The application provides objective video quality algorithms, such as PSNR, SSIM, and PW-SSIM and also incorporates a recently published technique for stereoscopic video quality assessment called Disparity Weighting (DW), which comprises the following algorithms: DPSNR, DSSIM and DPW-SSIM. Numerical results corresponding to the performance of the objective measurements, acquired using the proposed application, are presented. The application aims to contribute to the standardization and development of objective algorithms for stereoscopic content. As an open-source tool to be used by the academia and the industry, the application is used to evaluate impairments in stereoscopic video signals, caused by processing, compression and transmission techniques.

Journal of ICT, Vol. 2, 247–268.
doi: 10.13052/jicts2245-800X.233

Keywords: Stereoscopic Video, Video Quality Assessment, Objective Algorithms.

1 Introduction

The diversity of the digital multimedia content available in a wide range of services requires the use of different schemes to distribute the signals over a variety of digital transmission technologies. Furthermore, the multimedia streams must be adjusted to fulfill the requirements of applications, such as, IPTV, mobile TV, digital broadcasting, video on demand and video surveillance, which already include stereoscopic video content. It is important to consider the parameters that have a significant impact on the consumer electronics, such as, bandwidth, bit-rate, storage space and power consumption.

The visualization of stereoscopic video content requires more storage space and processing capacity from the devices. Moreover the power consumption is a critical parameter for mobile devices. Equally important are compression, transmission, coding, quantization and techniques to accommodate the multimedia signals, which can cause impairments that degrade the quality of the content. The Video Quality Assessment (VQA) is used to establish the performance of a video processing system and to obtain optimal parameters that maintain the Quality of Service (QoS) in a pre-determined condition [1]. Therefore, it is necessary to evaluate the quality of stereoscopic video signals in a fast, accurate and reliable way.

The approaches to evaluate the video quality can be subjective or objective. The subjective assessment involves psycho-visual experiments with humans, which provide adequate accuracy. The complete methodologies of the subjective experiments are described in Recommendations ITU-T P.910 [2] and ITU-R BT.500-13 [3]. However, this approach is time-consuming and less practical for real-time applications. On the other hand, the objective assessment is a fast and low cost alternative. Objective algorithms use statistical data of the video, combined with specific characteristics of the Human Visual System (HVS), to estimate the quality. It is classified according to the availability of the original signal as following: *full reference,* in which the original video is compared with the degraded video under test; *reduced reference,* whenever only characteristics of the original video are available for comparison with the video under test, and *no reference,* in which only the video under test is used for quality assessment.

Full-reference objective algorithms were developed to evaluate the 2D video quality, and were incorporated in Recommendation ITU-T J.144 [4]. Recent publications indicate that the perceptual characteristics increase the performance of the objective VQA algorithms [5–8]. However, stereoscopic video signals present the depth, which is a new component that can be considered in the design of the objective algorithms. The disparity has been used as a depth estimation in the development of objective algorithms [9, 10]. In addition, the recent development of objective stereoscopic video quality algorithms have encouraged the research in this subject [11–13], as well as the efforts to build standards for evaluation of subjective and objective stereoscopic video quality models [14].

This article presents a platform independent application, called Squales (Stereoscopic Video Quality Evaluation Software), with a graphical user interface for objective stereoscopic VQA. It was implemented using the Qt application framework and C++ programming language and can be used to implement important objective algorithms, such as, PSNR, SSIM, PW-SSIM, DPSNR, DSSIM, and DPW-SSIM. The performance of the objective algorithms was verified using the subjective data provided by the NAMA3DS1-COSPAD1 database [15, 16]. Statistical measures were used to compare the performance of the objective algorithms. Confidence intervals and hypothesis tests, for the Pearson correlation coefficient, are presented.

The article is organized as follows. Section 2 discusses applications. A review of the objective algorithms is presented in Section 3. Section 4 presents the Squales application. Numerical results, that validated the performance of the objective algorithms, are discussed in Section 5. Section 6 presents the conclusions and future work.

2 Related Applications

An application that implements image and video quality algorithms has a significant impact on academia and industry to enhance and to develop new video quality algorithms, as well as, to determine the performance of video processing techniques.

In particular, an image quality toolbox for MATLAB[1] ® was implemented by Sprljan [17]. This toolbox does not present a Graphical User Interface (GUI)

---

[1]MATLAB is a registered trademark under Copyright license with all rights reserved to MathWorks, Inc.

and the objective algorithms provided are limited to the error sensitivity approach, such as Mean Square Error (MSE), Peak Signal-to-Noise Ratio (PSNR), Normalized Absolute Error (NAE) and Average Difference (AD), which do not present good fitting with subjective scores.

Similarly, Gaubatz [18] provides a MATLAB package for image quality assessment. This application is command-line based, and its usage is not intuitive. Nevertheless, it provides image quality algorithms such as MSE, PSNR, SSIM, VIF (Visual Index Fidelity) and another algorithms based on the Human Visual System (HVS).

In the same way, a MATLAB-based framework for 2D image and video quality assessment was developed by Murthy and Karam [19]. This framework provides a GUI-based approach and it has various objective image and video quality algorithms, full-reference, reduced-reference and no-reference, and it supports different image and video formats. It also includes an interface for subjective evaluation, generation of simulated impairments, and correlation analysis between subjective and objective measures.

However, MATLAB is a high-cost application and time-consuming, for most devices, mainly to execute operations with large video files. MATLAB requires a robust hardware specification, which makes the analysis of stereoscopic video signals, in mobile devices scenario and real-time applications, almost impracticable.

Recently, Ucar *et al.* [20] proposed the Video Tester – a framework for video quality assessment over IP networks. It was implemented using Python programming language and it performs parameter extraction in packet, bitstream, and picture level of the video processing and transmission in order to gather information for quality evaluation. The Video Tester is a Linux application and it depends of Open Source Computer Vision Library (OpenCV), GStreamer, Matplotlib, and another libraries, that are not optimized for multi-platform compatibility.

3 Objective Evaluation Algorithms – A Review

3.1 Notation

Let $V = \{v_L(x, y, n), v_R(x, y, n)\}$ be a stereoscopic video signal, in which the scalar functions v_L and v_R correspond to left and right views respectively; such that $\{(x, y, n) \in \mathbb{Z}^3 : 1 \leq x \leq X; 1 \leq y \leq Y; 1 \leq n \leq N\}$, in which X, Y, and N represent the number of lines, columns, and frames respectively. This model was chosen because it is similar to that used in the standard ISO/IEC 14496-10 – MPEG-4 Part 10 (H.264/AVC) in Stereo and Multiview High Profiles.

Let F and H be a stereoscopic reference video signal and a stereoscopic video signal under test respectively. A full reference objective algorithm for stereoscopic video quality assessment is a function G, such that its outcome $(G(F, H))$ represents the quality of H with respect to F.

Further, the following definition is used in this paper

$$G(F, H) := \frac{G(f_L, h_L) + G(f_R, h_R)}{2}, \tag{1}$$

since the importance of left and right views for quality assessment is the same.

3.2 Peak Signal-to-Noise Ratio

Let $f(x, y, n)$ and $h(x, y, n)$ be scalar functions that represent 2D video signals. The Mean Square Error (MSE) between them is computed as

$$\text{MSE}(f, h) = \frac{\sum_{n=1}^{N} \sum_{x=1}^{X} \sum_{y=1}^{Y} [f(x, y, n) - h(x, y, n)]^2}{N \cdot X \cdot Y}. \tag{2}$$

The Peak Signal-to-Noise Ratio (PSNR) is computed as

$$\text{PSNR}(f, h) = 20 \cdot \log_{10} \left[\frac{\text{MAX}}{\sqrt{\text{MSE}(f, h)}} \right] \text{dB}, \tag{3}$$

in which $\text{MAX} = 2^b - 1$, b is the number of bits used in the quantization of the gray value scale, and $\text{MSE}(f, h)$ is the Mean Square Error between f and h. In this paper $b = 8$ was used.

3.3 Structural Similarity Index

The Structural SIMilarity (SSIM) [21] is a full-reference approach to image and video quality assessment based on the assumption that the HVS is highly adapted to recognize structural information in visual environments. Therefore, changes in structural information provide a good approximation to the quality perceived by the human visual system.

The SSIM(f, h) is computed as a product of three measures over the luminance plane: luminance comparison $l(f, h)$, contrast comparison $c(f, h)$ and structural comparison $s(f, h)$:

$$l(f, h) = \frac{2\mu_f \mu_h + C_1}{\mu_f^2 + \mu_h^2 + C_1}, \tag{4}$$

$$c(f, h) = \frac{2\sigma_f \sigma_h + C_2}{\sigma_f^2 + \sigma_h^2 + C_2}, \tag{5}$$

$$s(f, h) = \frac{\sigma_{fh} + C_3}{\sigma_f \sigma_h + C_3}, \tag{6}$$

in which μ is the sample mean, σ is the sample standard deviation, σ_{fh} is the covariance, $C_1 = (0.01 \cdot 255)^2$, $C_2 = (0.03 \cdot 255)^2$ and $C_3 = \frac{C_2}{2}$.

The structural similarity index is described as

$$\text{SSIM}(f, h) = [l(f, h)]^\alpha \cdot [c(f, h)]^\beta \cdot [s(f, h)]^\gamma, \tag{7}$$

in which usually $\alpha = \beta = \gamma = 1$.

In practice the SSIM is computed for an 8×8 sliding squared window or for an 11×11 Gaussian-circular window. The first approach was used in the experiments. For two videos, which are subdivided into J windows, the SSIM is computed as

$$\text{SSIM}(f, h) = \frac{1}{J} \sum_{j=1}^{J} \text{SSIM}(f_j, b_j), \tag{8}$$

in which f_j is the video signal observed in the j-th window.

3.4 Perceptual Weighted Structural Similarity Index

Regis et al. [22] proposed a technique called Perceptual Weighting (PW), which combines the local Spatial Perceptual Information (SI), as a visual attention estimator, with the SSIM, since experiments indicate that the quality perceived by the HVS is more sensitive for areas of intense visual attention [8].

The PW technique uses the local spatial perceptual information to weigh the most visually important regions. This weighting is obtained as follows: compute the magnitude of the gradient vectors in the original video using Sobel operator, then generate a perceptual map, in which the pixel values are the magnitude of the gradient vectors. The frame is partitioned into window 8×8 pixels, and the SI for each block is computed as

$$SI(f_j) = \sqrt{\frac{1}{K-1}\sum_{k=1}^{K}(\mu_j - |\nabla f_j(k)|)^2}, \qquad (9)$$

in which μ_j represents the sample average of the perceptual map in the j-th window, K is the number of gradient vectors in the j-th window and $|\nabla f_j(k)|$ is the magnitude of the k-th gradient vector in the j-th window. For the case the frames are partitioned uniformly in squares 8×8, $K = 64$.

Finally, the Perceptual Weighted Structural Similarity Index (PW–SSIM) is computed as

$$PW\text{--}SSIM(f, h) = \frac{\sum_{j=1}^{J} SSIM(f_j, h_j) \cdot SI(f_j)}{\sum_{j=1}^{J} SI(f_j)}. \qquad (10)$$

3.5 Disparity Weighting Technique

The disparity presented in a stereoscopic video signal is an information related to the sense of the stereo perception [9]. This information is computed as the difference between two corresponding pixels on the left and on the right views. Indeed, the disparity should be considered in the development of objective algorithms to improve the correlation between objective prediction and subjective scores.

The disparity map, D(F), is computed as

$$D(F(x, y, n)) := |f_L(x, y, n) - f_R(x, y, n)|, \quad \forall\, (x, y, n). \qquad (11)$$

Regis *et al.* [10] included the disparity information into objective algorithms by means of a weighted average of objective measurements with the values contained on disparity map. This approach was implemented into three objective algorithms, PSNR, SSIM, and PW-SSIM, developing DPSNR, DSSIM and DPW–SSIM.

The $DMSE_L$, i.e., DMSE for left view, is computed as

$$DMSE_L(F, H) =$$
$$\frac{\sum_{n=1}^{N}\sum_{x=1}^{X}\sum_{y=1}^{Y}[f_L(x, y, n) - h_L(x, y, n)]^2 \cdot D(F(x, y, n))}{\sum_{n=1}^{N}\sum_{x=1}^{X}\sum_{y=1}^{Y} D(F(x, y, n))} \qquad (12)$$

and the $DPSNR_L$ is computed as

$$DPSNR_L(F,H) = 20 \cdot \log_{10} \left[\frac{MAX}{\sqrt{DMSE_L(F,H)}} \right] \text{dB}. \qquad (13)$$

The DMSE and the DPSNR for right view ($DMSE_R$ and $DPSNR_R$) are computed in the same manner. Then the overall DPSNR is the average between $DPSNR_L$ and $DPSNR_R$.

The DSSIM is computed as

$$DSSIM(F,H) = \frac{\sum\limits_{j=1}^{J} SSIM(F_j, H_j) \cdot D(F_j)}{\sum\limits_{j=1}^{J} D(F_j)}, \qquad (14)$$

in which $D(F_j)$ is the average disparity in the j-th window.

The DPW–SSIM is computed as

$$DPW{-}SSIM(F,H) = \frac{\sum\limits_{j=1}^{J} SSIM(F_j, H_j) \cdot [SI(F_j) \cdot D(F_j)]}{\sum\limits_{j=1}^{J} [SI(F_j) \cdot D(F_j)]}. \qquad (15)$$

4 Proposed Application

This paper presents a platform independent application with an user-friendly GUI for objective stereoscopic VQA. The application was developed using C++ programming language with Qt[2] ® 4.8.4 (64 bits) under the license LGPL v2.1, using Qt Creator® 2.6.2 as an integrated development environment and GNU C and C++ Compiler (GCC v4.7.2).

Qt is a multi-platform application framework, developed in C++ programming language, which is widely utilized for developing applications with GUI. For instance, Qt is used in applications such as Google Earth and KDE. The C++ programming language was chosen because it is more efficient than other programming languages such as MATLAB. In fact, MATLAB is

---

[2]Qt and Qt Creator are registered trademarks under Copyright license with all rights reserved to Digia Plc.

time-consuming and requires large computational resources, which can be a problem for mobile devices. On the other hand, C++ is appropriate for mobile devices, because it is fast, even for limited computing resources.

The block diagram of the Squales application is depicted in Figure 1. The project architecture of the Squales application is shown in Figure 2, evidencing the facility to insert new objective algorithms into the project, which only requires their implementation in C++ programming language. Figure 3 depicts the GUI of the proposed application running on different platforms.

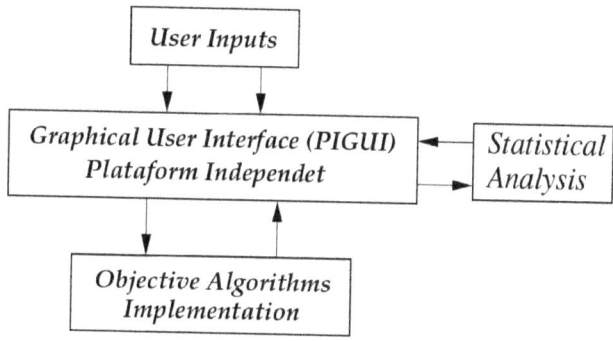

Figure 1   Squales block diagram

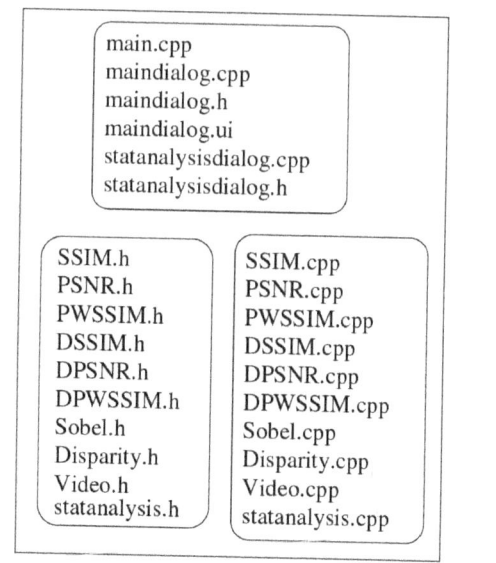

Figure 2   Squales project

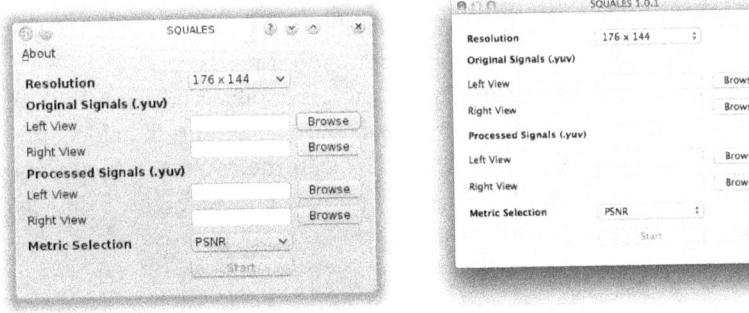

(a) OpenSUSE 12.3 KDE.   (b) Mac OSX

Figure 3   Main window of the Squales application

4.1 User Inputs

The application requires that the user indicates some input parameters, including: spatial resolution, video signals and metric selection.

1. *Spatial resolution*: The resolutions available are: 176×144 (QCIF); 352×288 (CIF); 640×480; 704×480; 720×480; 768×576; 872×480; 1024×576; 1048×576; 1280×720 and 1920×1080.
2. *Video signals*: The application supports video signals in YCbCr color space format, with 4:4:4, 4:2:2 and 4:2:0 chroma sub-sampling.
3. *Metric selection*: The metrics available are those discussed in the previous section, namely: PSNR, SSIMPW-SSIM, DPSNR, DSSIM, and DPW-SSIM.

4.2 Objective Algorithms Implementation

The implementation of the objective algorithms available in the proposed application was made in C++ programming language, according to the description presented in the previous section. The output of the objective algorithms is a text file. The time consumed, in seconds, by the objective algorithm is also presented.

4.3 Statistical Analysis Tool

In order to evaluate the performance of an objective algorithm, Squales provides a basic statistical analysis tool that computes the Pearson correlation coefficient between the sets of objective results and subjective scores, and

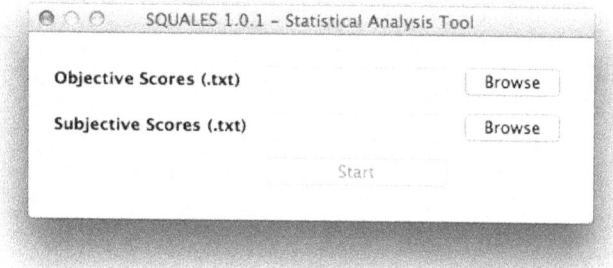

Figure 4   Squales statistical analysis tool

it also provides the fitted coefficients used to estimate a polynomial relation between those sets. In this sense, Squales requires that the user enter the sets of values in format of text files (.txt). It is important to highlight that the data in both files must be aligned. For instance, if an objective result correspondent to video#1 is in the first line of the text file of objective scores, then the subjective score correspondent to video#1 must be in the first line of the text file of subjective scores.

This statistical analysis tool was developed using Armadillo C++ Linear Algebra Library [23].

5 Numerical Results

The Squales application was validated using the stereoscopic video signals available from NAMA3DS1-COSPAD1 database. The NAMA3DS1-COSPAD1 stereoscopic video quality database [15, 16] provides subjective results, using the Absolute Category Rating with Hidden Reference (ACR-HR) method [2], considering scenarios restricted for coding and for spatial degradations including: H.264/AVC and JPEG2000 coding. Table 1 summarizes the coding conditions used in the subjective experiments performed by NAMA3DS1-COSPAD1 project.

5.1 Statistical Measures

The statistical measures used to compare the performance of the objective algorithms were: the Pearson Linear Correlation Coefficient (PLCC), the Spearman Rank-Order Correlation Coefficient (SROCC), the Kendall Rank Order Correlation Coefficient (KROCC) and the Root Mean Square Error

Table 1   Coding conditions used in NAMA3DS1-COSPAD1

Type	Parameter
Video Coding (H.264/AVC)	QP = 32
	QP = 38
	QP = 44
Still Image Coding (JPEG2000)	Bit Rate = 2 Mbits/s
	Bit Rate = 8 Mbits/s
	Bit Rate = 16 Mbits/s
	Bit Rate = 32 Mbits/s

(RMSE). In practice, PLCC evaluates the accuracy, SROCC and KROCC evaluate the monotonicity and the RMSE estimate the consistency of an objective model prediction with respect to human subjective scores available from NAMA3DS1-COSPAD1.

The statistical measures were computed after performing a non-linear regression on the objective video quality assessment algorithmic measures, using a four parameter monotonic cubic polynomial function to fit the objective prediction to the subjective quality scores. The function is as follows [4]

$$\text{DMOS}_l^{(p)} = \beta_1 + \beta_2 \cdot Q_l + \beta_3 \cdot Q_l^2 + \beta_4 \cdot Q_l^3, \tag{16}$$

in which Q_l represents the quality that an objective algorithm predicts for the l-th video in the NAMA3DS1-COSPAD1 Video Quality Database and $\text{DMOS}_l^{(p)}$ is the fitted objective score. The β coefficients are determined using a simple non-linear least squares optimization.

The results of the statistical measures, to compare the performance of the objective algorithms, are presented in Table 2. The two best results are shown in boldface. It is noted that the inclusion of the disparity weighting technique enhance the performance of the objective algorithms. The DPW–SSIM achieved the best performance for all scenarios, indicating that the perceptual weighting combined with the disparity weighting presented a significant role for the prediction of quality assigned by the HVS.

Figure 5 presents the trend between the set of subjective scores and the set of objective results. The scatter plots indicates a lower dispersion around of the prediction curve for the algorithms that use the disparity weighting.

Table 2 Performance measures of the objective algorithms

(a) H.264 scenario

Algorithm	PLCC	SROCC	KROCC	RMSE
PSNR	0.774946	0.721424	0.533869	0.689299
SSIM	0.730523	0.716222	0.555117	0.744770
PW-SSIM	0.915983	0.906776	0.756978	0.437573
DPSNR	0.863640	0.838604	0.640111	0.549789
DSSIM	0.901635	0.892266	0.746354	0.471688
DPW-SSIM	0.954403	0.937166	0.815412	0.325572

(b) JPEG2000 scenario

Algorithm	PLCC	SROCC	KROCC	RMSE
PSNR	0.828049	0.825865	0.662380	0.734844
SSIM	0.896314	0.907419	0.750010	0.581185
PW-SSIM	0.972477	0.965980	0.860836	0.305388
DPSNR	0.914034	0.927596	0.770629	0.531663
DSSIM	0.969310	0.962132	0.853104	0.322222
DPW-SSIM	0.975911	0.971048	0.865991	0.285951

(c) Joint scenario

Algorithm	PLCC	SROCC	KROCC	RMSE
PSNR	0.790152	0.766721	0.588923	0.750780
SSIM	0.832476	0.841566	0.658728	0.678694
PW-SSIM	0.951992	0.943427	0.800988	0.374981
DPSNR	0.875461	0.858578	0.678167	0.592001
DSSIM	0.944039	0.942530	0.801872	0.404026
DPW-SSIM	0.967001	0.955609	0.830147	0.312082

5.2 Hypothesis Tests for ρ

A statistical analysis was performed under the following hypothesis

$$\begin{cases} H_0 : \rho = \rho_0, \\ H_1 : \rho > \rho_0, \end{cases} \tag{17}$$

in order to verify whether the Pearson correlation coefficients have increased significantly.

Firstly, the Fisher's Transformation was applied

$$Z = \frac{1}{2} \log_e \left(\frac{1+r}{1-r} \right) = \operatorname{arctanh}(r), \tag{18}$$

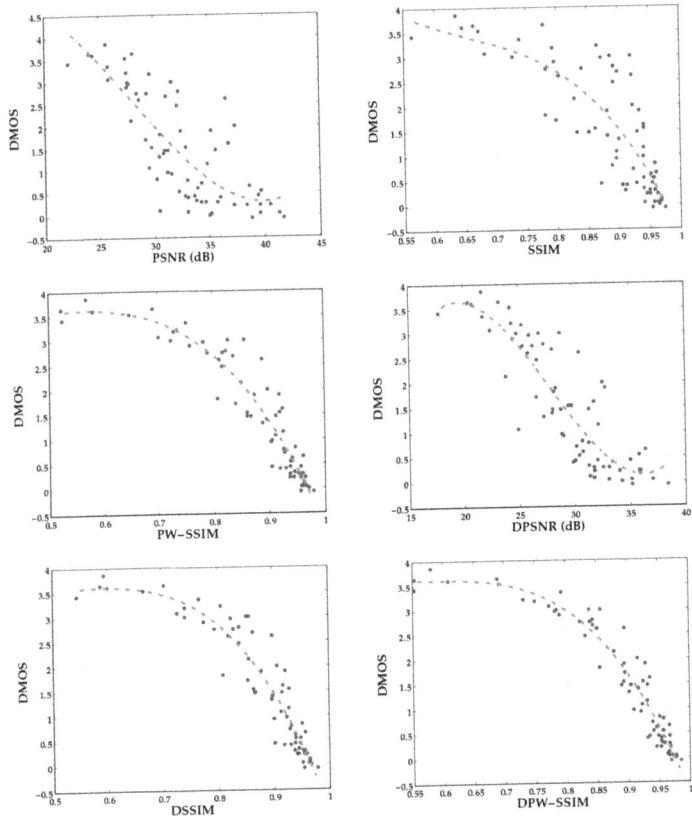

Figure 5   Scatter plots of subjective scores (DMOS) versus model prediction. Each sample point represents a 3D test video sample

in which r is the sample correlation coefficient (PLCC). Z follows approximately the Normal distribution $N(\mu_Z, \sigma_Z)$ with

$$\mu_Z = \text{arctanh}(\rho_0), \qquad \sigma_Z^2 = \frac{1}{\mathcal{N}_s - 3}. \tag{19}$$

Then, the Critical Region (CR) for Z, in the significance level of 95%, is

$$\text{CR} = \{Z : Z > \mu_Z + 1.654 \cdot \sigma_Z\} = \{Z : Z > Z_{CR}\}. \tag{20}$$

For each Pearson correlation coefficient presented (r) in Tables 2a, 2b and 2c, CR, and Z_0 were computed according to the Formulas (20) and (18). If $Z_0 \notin$ CR, the hypothesis $\mathcal{H}_0$ is accepted, i.e., there is not evidence that the Pearson correlation coefficient has increased, otherwise the hypothesis

Table 3 Hypothesis test for the correlation coefficient (H.264 – JPEG2000 – Joint Scenario)

Algorithms	PSNR	DPSNR	SSIM
PSNR	–	0 – 0 – 0	0 – 0 – 0
DPSNR	0 – 1 – 1	–	1 – 0 – 0
SSIM	0 – 0 – 0	0 – 0 – 0	–
DSSIM	1 – 1 – 1	0 – 1 – 1	1 – 1 – 1
PW-SSIM	1 – 1 – 1	0 – 1 – 1	1 – 1 – 1
DPW-SSIM	1 – 1 – 1	1 – 1 – 1	1 – 1 – 1

$\mathcal{H}_1$ is accepted meaning that the Pearson correlation coefficient has increased with a confidence level of 95%.

For instance, for $\rho_0 = 0.730523$ (PLCC for SSIM in H.264 scenario), then Z_{CR} is computed according to the Formula (20). The next step is to transform the another Pearson correlation coefficients in Z_i (the index i meaning the i-th algorithm) according the Formula (18). If $Z_i > Z_{CR}$ then the i-th algorithm presents a PLCC more significant than ρ_0 with a confidence level of 95%.

In Table 4 the values '0' and '1' mean that one of the hypothesis $\mathcal{H}_0$ or $\mathcal{H}_1$ was accepted. In practice, a symbol value of '1' indicates the statistical performance of the objective algorithm in the row is superior to that of the objective algorithm in the column. On the other hand, a symbol value of '0' suggests the statistical performance of the objective algorithm in the row is equivalent to that of the objective in the column. The sequence of the values in a cell corresponds to the hypothesis test for H.264, JPEG2000, and Joint scenarios respectively.

5.3 Confidence Interval for ρ

Figure 6 presents a 95% confidence interval for ρ in the H.264 and in the JPEG2000 scenarios under the hypothesis $\mathcal{H}_0 : \rho = 0$, $\mathcal{H}_1 : \rho \neq 0$. The Z Fisher's transformation was applied to r to compute the confidence interval.

Table 4 Hypothesis test for the correlation coefficient (H.264 – JPEG2000 – Joint Scenario)

Algorithms	DSSIM	PW-SSIM	DPW-SSIM
PSNR	0 – 0 – 0	0 – 0 – 0	0 – 0 – 0
DPSNR	0 – 0 – 0	0 – 0 – 0	0 – 0 – 0
SSIM	0 – 0 – 0	0 – 0 – 0	0 – 0 – 0
DSSIM	–	0 – 0 – 0	0 – 0 – 0
PW-SSIM	0 – 0 – 0	–	0 – 0 – 0
DPW-SSIM	1 – 0 – 1	0 – 0 – 0	–

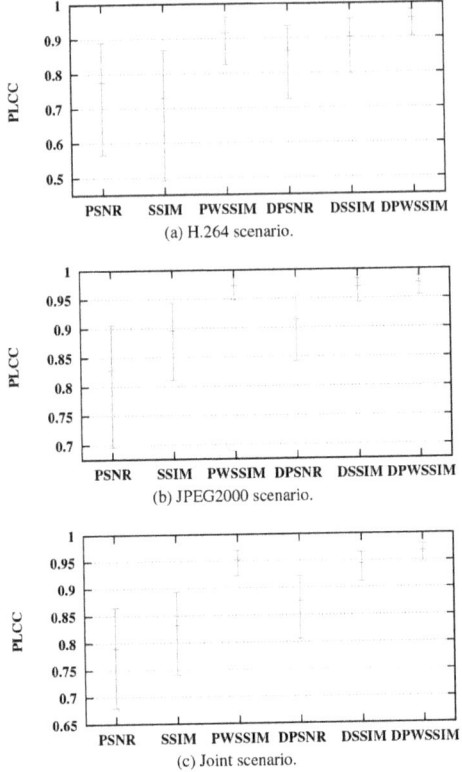

(a) H.264 scenario.

(b) JPEG2000 scenario.

(c) Joint scenario.

Figure 6   The 95% confidence intervals for the PLCC

Under that hypothesis, Z follows the Normal distribution with zero mean and with variance given by Formula (19). The confidence interval for this random variable is defined as

$$\text{IC}(z, 1 - \alpha) = (Z - z_{1-\alpha} \cdot \sigma_Z, \ Z + z_{1-\alpha} \cdot \sigma_Z). \tag{21}$$

For $\alpha = 0.05$, i.e., an interval with 95% of confidence, $z_{0.95} = 1.96$. and the Formula (21) may be rewritten as

$$\text{IC}(z, 0.95) = (Z - 1.96 \cdot \sigma_Z, \ Z + 1.96 \cdot \sigma_Z). \tag{22}$$

The inverse of the Z Fisher's transformation is

$$r = \frac{e^{2z} - 1}{e^{2z} + 1} = \tanh(z), \tag{23}$$

and the confidence interval in terms of r is defined as

$$IC(r, 0.95) = \left(\frac{e^{2 \cdot (Z - 1.96 \cdot \sigma z)} - 1}{e^{2 \cdot (Z - 1.96 \cdot \sigma z)} + 1}, \ \frac{e^{2 \cdot (Z + 1.96 \cdot \sigma z)} - 1}{e^{2 \cdot (Z + 1.96 \cdot \sigma z)} + 1} \right). \quad (24)$$

Figure 6 suggests that, besides an increase in the correlation coefficient, there is a reduction in the length of the PLCC confidence interval for the algorithms that use the disparity weighting technique.

6 Conclusions

An application that computes the stereoscopic video quality is important for the development and standardization of new objective stereoscopic video quality prediction models, and it also has a significant impact on academia and industry, because it can improve stereoscopic video processing techniques and stereoscopic video services. A platform independent application with GUI for objective stereoscopic video quality assessment was presented.

The main contributions of the application are: it is platform independent, it quickly computes the stereoscopic video quality, because the C++ programming language was used in the implementation, it presents a user-friendly GUI, and it presents a statistical analysis tool to compute the performance of objective algorithms. Squales is able to evaluate stereoscopic video signals in two-view model for several spatial resolutions with reliable objective algorithms specifically developed for stereoscopic video quality assessment.

The performance of the objective algorithms available in Squales was verified, using correlation coefficients, confidence interval for the PLCC and hypothesis test. These figures of merit validate the performance of the objective algorithms, evidencing the outstanding performance of the objective algorithms that include the disparity weighting technique, suggesting that Squales is an useful tool for be used by academia, industry, and standards organizations.

References

[1] K. Seshadrinathan, R. Soundararajan, A. C. Bovik, and L. K. Cormack. Study of Subjective and Objective Quality Assessment of Video. *IEEE Transactions on Image Processing*, pages 1427–1441, 2010.

[2] International Telecommunication Union. Recommendation ITU-T P.910: Subjective Video Quality Assessment Methods for Multimedia Applications. Technical report, ITU-T, 2008.

[3] International Telecommunication Union. Recommendation ITU-R BT.500-13: Methodology for the Subjective Assessment of the Quality of Television Pictures. Technical report, ITU-R, 2012.

[4] International Telecommunication Union. Recommendation ITU-T J.144: Objective Perceptual Video Quality Measurement Techniques for Digital Cable Television in the Presence of a Full Reference. Technical report, ITU-T, 2004.

[5] Z. Wang and Q. Li. Information Content Weighting for Perceptual Image Quality Assessment. *IEEE Transactions on Image Processing*, 20(5):1185–1198, 2011.

[6] C. D. M. Regis, J. V. M. Cardoso, and M. S. Alencar. Effect of Visual Attention Areas on the Objective Video Quality Assessment. In *Proceedings of the 18th Brazilian Symposium on Multimedia and the Web*, WebMedia '12, 2012.

[7] C. D. M. Regis, J. V. M. Cardoso, I. P. Oliveira, and M. S. Alencar. Performance of the Objective Video Quality Metrics with Perceptual Weighting Considering First and Second Order Differential Operators. In *Proceedings of the 18th Brazilian Symposium on Multimedia and the Web*, WebMedia '12, 2012.

[8] H. Liu and I. Heynderickx. Studying the Added Value of Visual Attention in Objective Image Quality Metrics Based on Eye Movement Data. In *16th IEEE International Conference on Image Processing*, 2009.

[9] A. Benoit, P. Le Callet, P. Campisi, and R. Cousseau. Quality Assessment of Stereoscopic Images. *EURASIP Journal on Image and Video Processing*, 2008(1):659024.

[10] C. D. M. Regis, J. V. M. Cardoso, I. P. Oliveira, and M. S. Alencar. Objective Estimation of 3D Video Quality: A Disparity-based Weighting Strategy. In *Proceedings of IEEE International Symposium on Broadband Multimedia Systems and Broadcasting (BMSB'13)*, 2013.

[11] J. Han, T. Jiang, and S. Ma. Stereoscopic Video Quality Assessment Model Based on Spatial-temporal Structural Information. In *IEEE Visual Communications and Image Processing (VCIP)*, 2012.

[12] D. Kim, D. Min, J. Oh, S. Jeon, and K. Sohn. Depth Map Quality Metric for Three-dimensional Video. In *Proceedings of XX SPIE Stereoscopic Displays and Applications*, volume 7237, 2009.

[13] L. Jin, A. Boev, A. Gotchev, and K. Egiazarian. 3D-DCT Based Perceptual Quality Assessment of Stereo Video. In *18th IEEE International Conference on Image Processing (ICIP)*, 2011.

[14] Video Quality Experts Group. Evaluation of Video Quality Models for use with Stereoscopic Three-Dimensional Television Content. Technical report, VQEG, 2012.

[15] IRCCyN-IVC. Nantes-Madrid 3D Stereoscopic Database. http://www.irccyn.ec-nantes.fr/spip.php?article1052,2012.

[16] M. Urvoy, M. Barkowsky, R. Cousseau, Y. Koudota, V. Ricorde, P. Le Callet, J. Gutierrez, and N. Garcia. NAMA3DS1-COSPAD1: Subjective Video Quality Assessment Database on Coding Conditions Introducing Freely Available High Quality 3D Stereoscopic Sequences. In *Quality of Multimedia Experience (QoMEX), 2012 Fourth International Workshop on*, 2012.

[17] N. Sprljan. MATLAB XYZ Toolbox, 2012. http://www.sprljan.com/nikola/matlab.

[18] M. Gaubatz. MeTriX MuX Visual Quality Assessment Package, 2007. http://foulard.ece.cornell.edu/gaubatz/metrix_mux/.

[19] A. V. Murthy and L. J. Karam. A MATLAB-based Framework for Image and Video Quality Evaluation. In *Second International Workshop on Quality of Multimedia Experience (QoMEX)*, 2010.

[20] I. Ucar, J. Navarro-Ortiz, P. Ameigeiras, and J. M. Lopez-Soler. Video Tester – A Multiple-metric Framework for Video Quality Assessment Over IP Networks. In *IEEE International Symposium on Broadband Multimedia Systems and Broadcasting (BMSB)*, 2012.

[21] Z. Wang, A. C. Bovik, H. R. Sheikh, and E. P. Simoncelli. Image Quality assessment: From Error Visibility to Structural Similarity. *IEEE Transactions on Image Processing*, 13(4):600–612, April 2004.

[22] C. D. M. Regis, J. V. M. Cardoso, and M. S. Alencar. Video Quality Assessment Based on the Effect of the Estimation of the Spatial Perceptual Information. In *Proceedings of 30th Brazilian Symposium of Telecommunications (SBrT'12)*, 2012.

[23] C. Sanderson. Armadillo: C++ linear algebra library, 2014. http://arma.sourceforge.net.

Biographies

J. Vinícius de Miranda Cardoso was born in Campina Grande, Brazil in 1992. He is a senior-year student of Electrical Engineering at the Federal University of Campina Grande (UFCG), Brazil. Currently, he was awarded with a scholarship sponsored by the Brazilian Government to pursue his studies in the US. Since his first year in college, he has been engaged with research projects related to image and video processing. Such works were published in reputed conferences such as IEEE BMSB'13, IEEE BMSB'14, IEEE SSIAI'14, and ITU Kaleidoscope 2014. The paper published in the ITU Kaleidoscope 2014 was nominated for the Best Paper Awarded (one of the best six papers) and, in that occasion, he was awarded with an Young Author Recognition by ITU. He is a student member of the IEEE, Brazilian Telecommunications Society (SBrT), and Institute for Advanced Studies in Communications (Iecom).

C. Danilo Miranda Regis was born in Guarabira, Brazil. He received his Bachelor Degree in Electrical Engineering, in 2007, his Masters Degree, in 2009, and his Doctor Degree, in 2013, both in Electrical Engineering, from the Federal University of Campina Grande (UFCG), Brazil. He is with the Iecom Executive Staff of the Journal of Communication and Information Systems (JCIS), since 2006. Since 2009, he is a professor at the Federal Institute of Education, Science and Technology of Paraba (IFPB), Brazil. His current interests include video quality metrics, video processing, multimedia, Digital TV, mobile TV, video transmission and biomedical engineering.

M. Sampaio de Alencar was born in Serrita, Brazil in 1957. He received his Bachelor Degree in Electrical Engineering, from Universidade Federal de Pernambuco (UFPE), Brazil, 1980, his Master Degree in Electrical Engineering, from Universidade Federal da Paraiba (UFPB), Brazil, 1988 and his Ph.D. from University of Waterloo, Department of Electrical and Computer Engineering, Canada, 1994. He is Chair Professor at the Department of Electrical Engineering, Federal University of Campina Grande, Brazil, founder and President of the Institute for Advanced Studies in Communications (Iecom). He published over 350 engineering and scientific papers and 16 books. He is Emeritus Member of the SBrT. He is a Registered Professional Engineer and a columnist for the traditional Brazilian newspaper Jornal do Commercio. He is Vice-President External Relations of SBrT.

On Data Program Interfaces

D. Namiot[1] and M. Sneps-Sneppe[2]

[1]*Lomonosov Moscow State University*
[2]*Ventspils University College*
E-mail: {dnamiot; manfreds.sneps}@gmail.com

Received: 24 November 2014;
Accepted: 9 March 2015

Abstract

In this paper, we discuss the global unified standards for the Internet of Things software products and existing de-facto standards (practical approaches). Using specific examples of interaction with Bluetooth Low Energy tags we compared existing approaches to the development and the proposed global standards (FI-WARE). Can the proposed unified approach to the creation of services to cover all the possible use cases and scenarios for Internet of Things and Smart Cities services? The paper emphasizes the need to address the simplicity of the development and highlights the critical importance of such a thing as the time to market for new applications and services. By our opinion, time to market and the simplicity of the development are key parameters for any proposed software standard.

Keywords: Internet of Things, FI-WARE, API, DPI, Bluetooth.

1 Introduction

This paper presents an extended version of our presentation for ITU Kaleidoscope Conference [1]. In this article we want to compare unified programming standards and practices of development.

In our research, we will talk about the standards that affect software development. And even more precisely, we will discuss Internet of Things (IoT) software. Nowadays, we can see many attempts to create a unified

Journal of ICT, Vol. 2, 269–288.
doi: 10.13052/jicts2245-800X.234

interface (platform, API) for IoT [2, 3]. So, it is correct to talk about an adaptation (adoption) of proposed standards by the existing development community. Only the adoption by the community ensures their wide distribution and use. Otherwise (which is not uncommon), we are faced with a situation where standards exist in parallel and independently of the established practice. In recent history, we can recall examples of real opposition to the proposed standard from the existed approaches and practices. For example, we can point out here the confrontation of TCP/IP protocol stack and the ISO, Corba and Web services, IIOP and XML, and on the same Web Services versus REST, XML versus JSON, etc [4]. And it is not always the confrontation ended with the benefit (in favor) of an official standard. As seems to us, the standards proposed for the software development have their own specifics.

Using specific examples of interaction with Bluetooth Low Energy tags, we compared the existing approaches to the development and the proposed global standards (FI-WARE). We would like to show that in many cases the proposed approaches are over-engineered and, actually, complicates the development process.

The rest of the paper is organized as follows. Section 2 describes the philosophy behind the software standards. Section 3 presents Bluetooth Low Energy tags. Section 4 discusses de-facto standard programming approach for Bluetooth Low Energy tags. Section 5 describes FI-WARE project. Section 6 targets Smart City programming in FI-WARE. Section 7 presents a discussion.

2 On Software Standards

In general, the typical standards committee starts with an idea. The idea could be adopted as a work item and then committee takes it through the successive stages of standardization (i.e. the standards process) [5]. As a result, we can get the standard specification. It could be implemented in a product or a service (a standard implementation). The paper [5] highlights the following sources of implementation problems:

- The idea that underlies a standard may not be implementable (e.g. too comprehensive).
- The ideal of consensus decision-making may affect the standards process. Typically, it leads to too many options ("a camel is a horse designed by a committee"). It affects, of course, the implementability of standards.
- Different use of terminology in a standard specification may lead to problems of interpretation, implementation and interoperability.

- Modest user requirements and cost-constraints in the implementation process may lead to partial standard compliance and incompatible implementations.

In the telecommunications world, the most interesting example is, of course, the whole story behind the Parlay. We saw a whole family of APIs: Parlay/OSA, Parlay X, which can be described as a simplified version of Parlay/OSA, then JAIN. This constant redesigning and repositioning of standards leads to a loss of meaning as to that constitutes a standard [6]. Parlay is also a great example of incompatibility for standards implementation.

An interesting fact is that for software systems the standard implementation does not oppose to some already well-established technology. With programming standards, where emerging de facto procedures play the major role, the whole picture is more complicated. At the time of introduction (during initial phases) common standards for their respective approaches to development have yet to develop. So, we can say that standard committee and developers outside of it are in the same conditions. But nevertheless, a more pragmatic approach invariably won. Each standard (relating to software development) proposes an all or nothing approach. And, accordingly, it sought to cover all the imaginable aspects. It leads (especially, with the above mentioned ideal of consensus decision-making) to the highest possible level of generalization. This, of course, is both the strength and weakness of this approach. In general, the problems with such an approach are the same as with the introduction new enterprise architecture.

What are the typical key challenges for the enterprise for implementing a new architecture? On the first hand it is the alignment of the organization's capabilities and strategy with the environment in which it operates. The same is true for the world of software development. A new standard will play a significant role in realigning the organizational structure for the software companies. Obviously, it changes the stability. And too many changes in a short time frame can cause the resistance. But in the same time misalignment can create big problems for software companies.

Capacity varies across development teams. So, the need for resource commitments is another reason for the resistance. Obviously, the resource commitments affect the implementation plan in any software company (team of developers). Other factors we can mention here are diversity of software companies, complexity of new offering as well as the need for coordination during the implementation phase.

But from our point of view, the main (determining) parameter is the answer to the main question of interest to developers. This issue is the time it takes to build services (applications) using a novel approach. Time is a key factor in software development [7]. Can we save a time with new approach? The biggest problem with Parlay was the fact that actually this standard increased the time for development (time to market for new services).

In light of the above, we want to compare two approaches to the development of services for the same type of sensors. One of these approaches is the proposed standard FI-WARE and the second one is the de facto folding (forming) approach to development: Bluetooth Low Energy (BLE). One of the approaches is pushed by the international community, where the second is initiated by one vendor. Actually, this use case is a typical for the software development. The most successful outcome for such kind of projects is when the private software development (software product) is becoming a standard, supported by the community.

We choose BLE related development as the typical example of sensors connected projects. From the one side, this development should be effectively supported by the proposed standard. But in the same time, we have a powerful vendor (Apple) with the own development. And this vendor has got enough power to turn any of the own projects into standard entity.

In general, we can describe two approaches for the software standards. Firstly, we can talk about some unified (generic) API. Any generic approach will cover a wide area of elements (devices and their models in case of IoT) and lets programmers use the same methods across the whole project. And the second approach is a specialized (vertical) API targets one particular class of entities (devices in case of IoT). Usually, the direct proposals from hardware vendors are more compact. Although of course, we cannot expect the broad applicability here.

3 Bluetooth Low Energy

Bluetooth low energy (BLE) as a technology started as a project in the Nokia Research Centre. Later it was adopted by the Bluetooth Special Interest Group [8]. The aim of this technology is to enable power sensitive devices to be permanently connected to the Internet. BLE sensor devices are typically required to operate for many years without needing a new battery [9]. As an ecosystem of other devices, BLE may also be known as Bluetooth v4.0 and is part of the public Bluetooth specification [10]. A device that operates Bluetooth v4.0 may not necessarily implement other versions of Bluetooth, in such cases it is known as a single mode device. Bluetooth classes are:

Class 1 (100 mW, 100 m range)

Class 2 (2.5 mW, 10 m range)

Class 3 (1 mW, 1 m range)

There are other low power technologies that could be deployed. For example, ANT which operates in the 2.4 GHz spectrum, ZigBee, Zigbee Reduced Function Device (RFD), etc [11].

Apple has been embedding Bluetooth Low Energy in its devices since iPhone 4s. Since iOS7 release, Apple has released iBeacon API. It is programming interface to low energy sensors from Apple (iBeacons).

In 2013, over 250 million Bluetooth Smart accessories are expected to ship. By 2016, this is expected to reach over a billion [12]. At this moment BLE has been supported by 100% of Apple iOS devices since iPhone 4S (including the iPod Touch, iPads, and all phones), and is roughly supported by 30% of new shipped Android phone models. So, these Bluetooth-enabled devices will interact with iOS and Android devices. Beacons can take any form factor and can be placed anywhere. From a developer perspective, they simply advertise data in peripheral mode by broadcasting some unique identifier. Actually, each tag broadcasts 3 numbers: own unique ID and two digits (so called minor and major). Minor and major could be configured in application in order to distinguish different areas. Application developers then use these numbers to understand the location of a particular device and connect an application (a mobile user) to a service or to content in the cloud.

Existing use cases include retailers and venue owners (see Figure 1).

Actually, the same picture is correct for Gimbal beacons [13] too, for example. As we can see, application to sensors communication link is a read-only channel. It is a very important remark. In practice, most of IoT applications are reading data only. It means, that the word API borrowed from telecom is very often means over-engineering here. Actually, most of the devices (sensors) do not accept data and cannot execute commands. They can only transmit data. They can do that by the own initiative (push) or do it as a response to any request (poll). This means also that the security constraints can be seriously reconsidered. Usually, by the obvious reasons, read only is much more safe, than read/write. Especially, if we keep in mind the fact, that present data for the reading is the main function for the most of our devices (sensors). They are not made for data hiding.

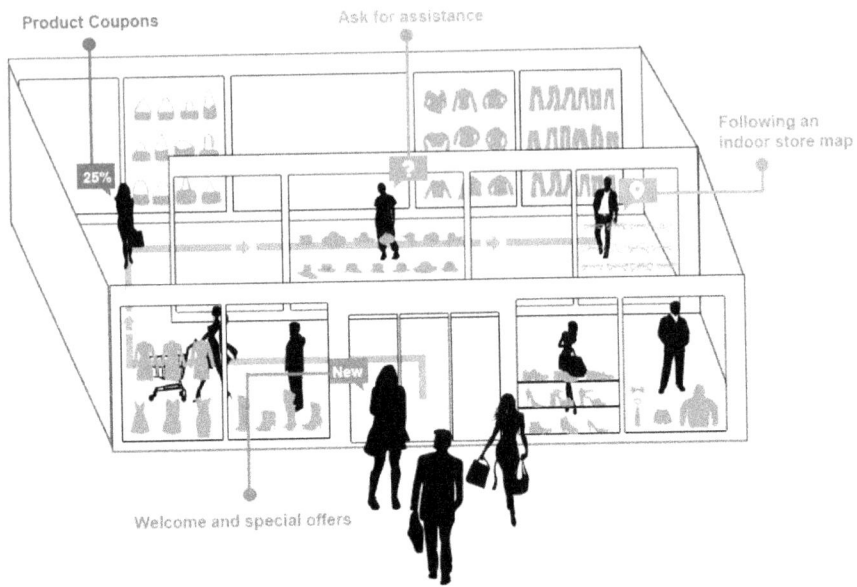

Figure 1   iBeacons use cases [14]

So, we can treat most of the devices as tags. We have presented iBeacon here just due to fact that the functionality described in the specific embodiments is wide enough by itself.

And the model, selected for this family of tags can be used in many applications. Therefore, it is important to standardize access to such facilities (tags). This standard should affect applications (services) that work with tags. Standardization will ensure the porting of applications and easy replacement of hardware devices.

And it is a perfect example for standard versus de facto approach in software development. On the one hand, we have FI-WARE standards that should cover sensors, on the other hand, we have developers around the existing mobile operating systems.

Bluetooth Low Energy uses Generic Attribute Profile (GATT) for service discovery [15]. One device looks for others and sends ID Packets. One or more other devices listen (these devices are discoverable). When acceptable packet is received, searching device may continue with this device or continue looking for other devices. When a desired device is found, connection process begins. There is always master-slave communication. One master controls communications and can support up to 7 active slaves. This system does not support slave-to-slave communication. So, only one device always initiates the connection. Other device has to be willing to accept a connection. Once listening device hears the ID packet with its address, it replies.

GATT provides a framework for developing profiles. A profile is composed of one or more services. A service is composed of characteristics or references to other services. Each characteristic contains a value and may contain optional information about the value [16]. Note, that it is a classical client-server model. Also, we should note that response's description is a good fit for JSON (one of the favorite formats for the modern programming).

Mobile phone supports the central role and BLE device supports the peripheral role. Once the phone and the BLE device established a connection, they start transferring GATT metadata to one another.

4 BLE Programming

The description below targets Android 4.3 (API Level 18). The *BluetoothAdapter* is required for any and all Bluetooth activity [17]. The *BluetoothAdapter* represents the device's own Bluetooth adapter. There is one Bluetooth adapter for the entire system.

```
// Initializes Bluetooth adapter.

final BluetoothManager bluetoothManager =
        (BluetoothManager)
getSystemService(Context.BLUETOOTH_SERVICE);

mBluetoothAdapter = bluetoothManager.getAdapter();
```

In order to find BLE devices, we can use the *startLeScan()* method. This method takes a *BluetoothAdapter.LeScanCallback* as a parameter. In this callback we can deal with scan results.

```
mBluetoothAdapter.startLeScan(mLeScanCallback);
```

Here is an implementation for callback as per Google manual:

```
private LeDeviceListAdapter mLeDeviceListAdapter;
```

. . .

```
// Device scan callback.
```

```
private BluetoothAdapter.LeScanCallback mLeScanCallback =
    new BluetoothAdapter.LeScanCallback() {
    @Override
    public void onLeScan(final BluetoothDevice device, int rssi,
        byte[] scanRecord) {
        runOnUiThread(new Runnable() {
        @Override
        public void run() {
            mLeDeviceListAdapter.addDevice(device);
            mLeDeviceListAdapter.notifyDataSetChanged();
        }
        });
    }
};
```

As soon as the device is updated, we can connect to a GATT server on the BLE device: *connectGatt()* method. This method takes three parameters: a Context object, *autoConnect* (a boolean indicating whether to automatically connect to the BLE device as soon as it becomes available), and a reference to a new callback.

Once our Android application has connected to a GATT server and discovered services, it can read and write attributes. So, if we skip the Android callback-based architecture, it is very straightforward: discover the device, make a connection and read attributes.

The application can ask to be notified when a particular characteristic changes on the device. And of course, we can stop and start scan.

Some vendors can provide a high level API for access to low energy tags (Samsung, HTC, Estimote [18]). There is *BeaconManeger* object (in iOS) that plays a role of proxy and hides details of the connection:

```
(void)beaconManager:(ESTBeaconManager *)manager
    didDetermineState:(CLRegionState)state
```

```
            forRegion:(ESTBeaconRegion *)region
{

   if(state == CLRegionStateInside)
   {

      [self setProductImage];

   }

   else

   {

      [self setDiscountImage];

   }

}
```

The programming pattern in both cases is the same. We need to find (address) our device (devices) and define a callback method for accepting data. We do not need even to make periodical requests. In this case, tags pushed data by themselves.

5 FI-WARE Approach

FI-WARE approach [19] is our example of unified API. The high-level goal of the FI-WARE project is to build the Core Platform of the Future Internet. FI-WARE is based upon elements (called Generic Enablers) which offer reusable and commonly shared functions serving a multiplicity of Usage Areas across various sectors [20]. Generic Enablers (GEs) are considered as software modules that offer various functionalitiesalong with protocols and interfaces for operation and communication. The Core Platform to be provided by the FI-WARE project is based on GEs linked to the following main FI-WARE Technical Chapters:

Cloud Hosting – the fundamental layer which provides the computation, storage and network resources, upon which services are provisioned and managed.

Data/Context Management – the facilities for effective accessing, processing, and analyzing massive volume of data, transforming them into valuable knowledge available to applications.

Applications/Services Ecosystem and Delivery Framework – the infrastructure to create, publish, manage and consume FI services across their life cycle, addressing all technical and business aspects.

Internet of Things (IoT) Services Enablement – the bridge whereby FI services interface and leverage the ubiquity of heterogeneous, resource-constrained devices in the Internet of Things.

Interface to Networks and Devices (I2ND) – open interfaces to networks and devices, providing the connectivity needs of services delivered across the platform.

Security – the mechanisms which ensure that the delivery and usage of the services is trustworthy and meets security and privacy requirements.

Let us see what is offered by IoT and I2ND enablers. From a physical architecture standpoint, IoT GEs have been spread in two different domains: Gateway and Backend.

The deployment of the architecture of the IoT Service Enablement chapter is typically distributed across a large number of Devices, several Gateways and the Backend (Figure 2).

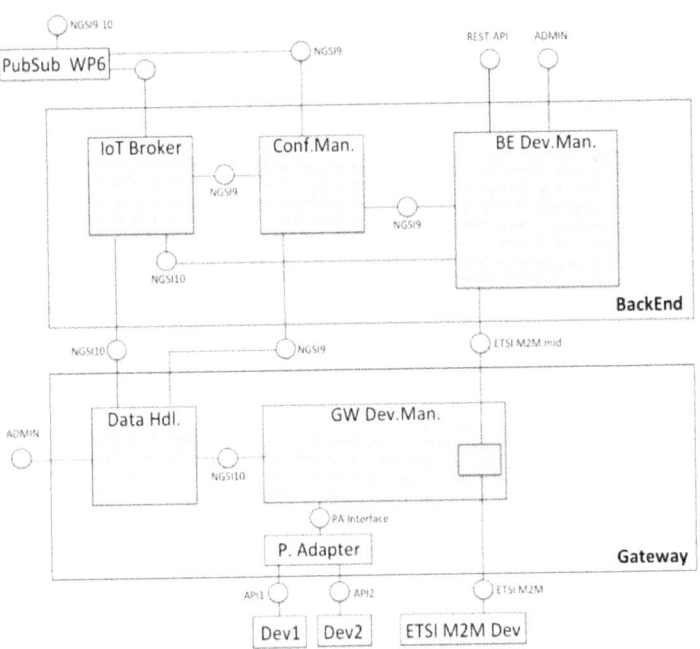

Figure 2   IoT service enablement

FI-WARE IoT Gateway is a hardware device that hosts a number of features of one or several Gateway Generic Enablers of the IoT Service Enablement. It is optional element and could be eliminated (as per FI-WARE spec). FI-WARE IoT Backend is a setting in the cloud that hosts a number of features of one or several Generic Enablers of the IoT Service Enablement. In the FI-WARE IoT model, at least one IoT Backend is mandatory, which will be connected to all IoT end devices either via IoT Gateway(s) and/or straight interfaces [19].

As we see, the unified approach makes data sharing (IoT Backend) mandatory. It could be one of the problems for the above-mentioned applications for access to BLE tags. For example, sharing data to cloud environment immediately adds security requirements. Also, it adds the complexity (read – increases price) for the technical installation. Obviously, it is much easier to install independent tags, rather than provide any cloud connectivity for each of them. Cloud based solution could be simple more costly. So, it is an absolute correct question: do we really need it for all imaginable scenarios? The typical use case for BLE tags is their inspection from smartphones (other personal devices). Smartphone (metering device) itself plays a role of gateway. Mandatory data sharing just adds the complexity.

6 Enablers for Sensing

According to [21], there are more than 60 FI-WARE Generic Enablers (GE) as common building blocks across Use Case projects, and more than 100 Specific Enablers as dedicated building blocks coming from the Use Case projects so as to support their proof of concept and build prototypes. Figure 3 shows GE for IoT:

Of course, this GE is supposed to work with many devices, so it is presented as much abstract as it is possible. For any particular preselected set of end-devices (sensors) the picture could be much simpler, of course.

The IoT Gateway and IoT Backend communicate with each other through an open standardized communication interface. The IoT gateway is responsible also for a protocol adaptation and control service. The protocol adaptation service will handle communication with different devices, while the control service will contain intelligent control logic that will deal with differences between the various implementations of the Gateway and access technologies, as shown in Figure 4.

Because it is a generic API, it provides (defines) all imaginable functions:

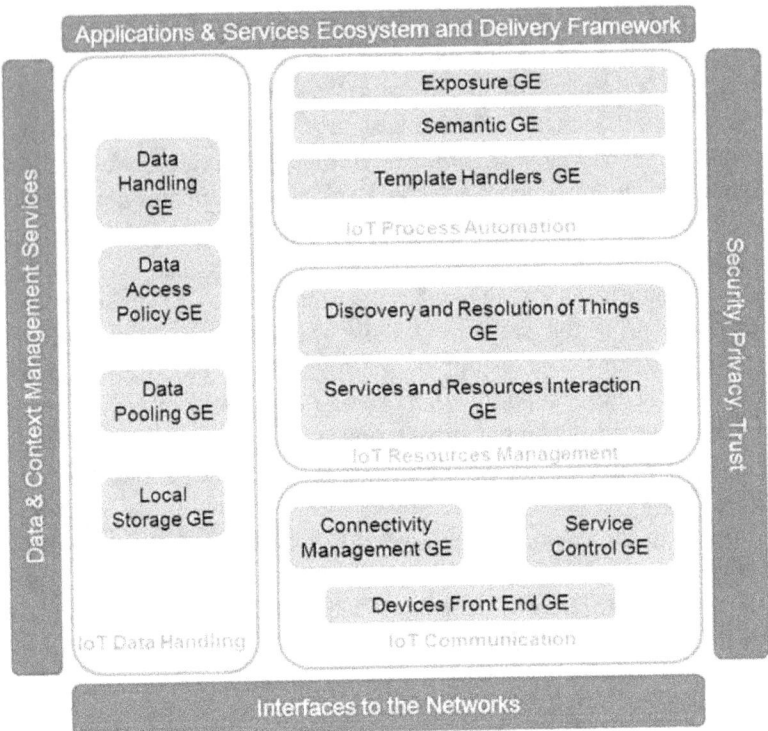

Figure 3 IoT GE

Communication protocols abstraction: a mechanism to enable unified communications between the IoT resources and the IoT backend.

Communication service capabilities: identification of and communication to IoT resources (identification of an IoT resource includes the mapping of physical network addresses to the IoT resource identifier).

Managed connectivity: definition of the interfaces towards the networks for the management of the connectivity.

Discontinuous connectivity: management of the IoT resources that are not always-on.

Support of the IoT mobility for the IoT resources that may physically move, or may change the access network.

Session management: management of the communication sessions to support mechanisms to handle reliability associated with network connections.

Traffic flow management: development of the mechanisms to deal with abnormal and occasional traffic conditions (e.g. overload traffic conditions).

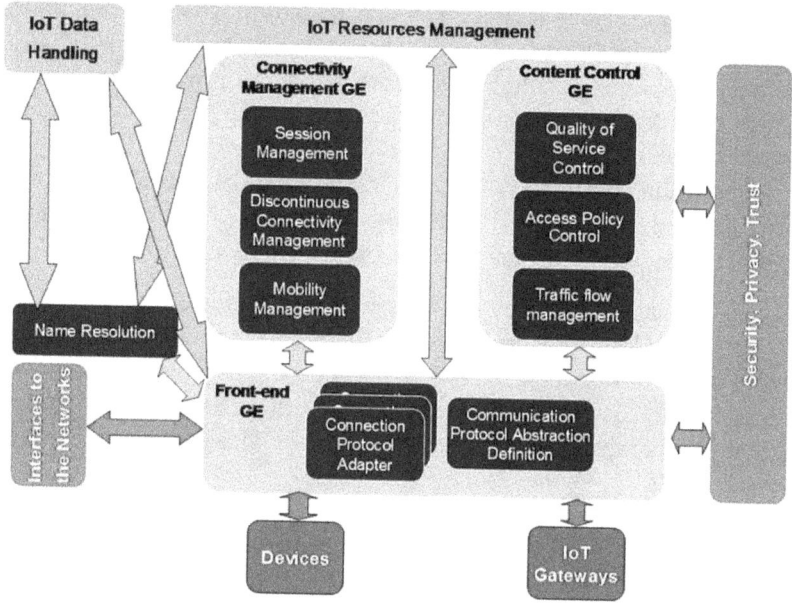

Figure 4 IoT communications [22]

It is interesting, that Fi-WARE documents mention so called "Data as a Service" component. FI-WARE assumes that often the data needs to be stored and later accessed by applications and / or by the IoT Services for its proper operation and management of IoT resources. So, "Data as a Service" is, firstly, some storage in their vision. This storage should follow to some structured approach and has to support consistency during time-to-time synchronization or sporadic events. And "Data as a Service" is, technically, some API for this storage.

We think, that "Data as a Service" should be a main function. At least, it is so for the vast majority of IoT systems. Almost all sensor-based systems fall into this category.

7 Discussion

In paper [23] we analyzed the typical IoT applications for wireless tags (e.g., iBeacons). We can present the top level definition as this:

- There is a set of sensors we need to poll periodically for getting new measurements

- There is a set of sensors we need to accept data from (push data – sensors initiated communications)
- The business process could be presented as a set of productions (rules). Each of the rules depends on some available data and, probably, on some global variables (states).
- The data availability always assumes the presence of data for any finite set of timestamps. In other words, the application makes conclusions (actions) depending on some window of measurements.

So, the top level architecture is very simple. The application finds a tag (device), obtains a descriptor and presents (defines) a callback function for getting data. It looks like a classical model for AJAX deployment: get data via some asynchronous call and proceed them in the predefined callback. It means that this approach will be supported sooner or later in some JavaScript environment.

The best candidate for top level presentation is Web Intents [24]. It is practically the same proxy for asynchronous calls accessible from JavaScript. Note, that the same approach works for many sensors. For example, our own SpotEx approach [25] that collects information from network nodes could be presented in the similar manner. So, as seems to us the following question is correct: do we really need Application Program Interfaces (API) always, or our goal could be described as Data Program Interfaces (DPI)?

We can describe DPI as an interface at the edge of an IoT device that exposes and consumes data. IoT devices very often do not support commands (instructions). Many of sensors just provide some data and nothing more. The above-mentioned "Data as a Service" approach is a good example. We could have data layer (probably, with the persistence support) and client-side processing (JavaScript).

Actually, we can mention several IoT-related projects with the similar conception. There is a Webinos project [26] with a very similar model. It is a European Union project, which aims at developing software components for the future Internet, in the form of Web Runtime Extensions. We can mention also OpenRemote project [27]. Another example is a MQTT protocol from IBM [28].

Actually, publish/subscribe systems are good examples of data centric communications. In the publish/subscribe communication model components which are interested in obtaining data register their interest. For our BLE example, they scan and discover tags. In publish/subscribe systems it is called subscription. Components, which want to share some information, do so by publishing their data. There is also a broker - the entity which ensures that the data gets from the publishers to the broker. This actor coordinates subscriptions. In our case this broker is personalized. It is an in-application proxy. There are three principal types of pub/sub systems: topic-based, type-based and content-based [29]. By topic-based systems, the list of topics is usually known in advance. E.g., in our case we know the tags and their data. In type-based systems, a subscriber states the type of data it is interested in (e.g., temperature data). It means that we need to maintain some directory of tags with metadata. In content-based systems the subscriber describes the content of messages it wants to receive. Subscriber describes types of data for receiving and condition for the delivery (e.g., the temperature is below a certain threshold). Publish/subscribe it is a good example for DPI too.

We can mention here a wide set of papers devoted to the Web of Sensors with linked data and HTTP based REST protocol [30]. But the basic point for such models could be a serious limitation. The models assume that the HTTP server is deployed on each sensor node, making it an independent and autonomous Web device.

Keeping in mind the above-mentioned examples with BLE tags, FI-WARE GE for this task should be personalized by some way and placed right on the mobile phone. For the many exiting BLE deployments backend is the mobile phone. And there is no cloud involved in this operation.

In general, the main question in our discussion could be provided in this form: can we state that excessive generalization (unification) in software standards could be a biggest source of the problems?

And it is a very significant question to the standard committee. Actually, this is a very important (if not a most important) point. It belongs to any standard devoted to the software development. Shall the standard follow to the "all or nothing" model and covers all the areas of life cycle? Actually, we are talking about all the imaginable areas at the moment of the standard's development. Note, that the above-mentioned scenario with direct access from mobile phone to tags (sensors) is quite common. And we should make its implementation by the most convenient way for mobile developers. They will be responsible for the putting new services in place.

The alternative approach is to target individual technical areas and let developers assemble applications like mashups from standard components. The key moment here is exactly time to market. Mashups allow developers seriously decrease time to market for new services. In the modern software architecture world, we can mention also micro-services approach [31].

We should mention in this context Data-as-a-Service (DaaS) approach [32]. DaaS (in its technical aspects) Data as a Service (DaaS) is an information provision and distribution model in which data files (including text, images, sounds, and videos) are made available to customers over a network. The key moment (again, we do not discuss the business model) is the separation for data and proceedings. The key delivery element is a data chunk rather than some API with the predefined model for data processing. Benefits of DaaS include outsourcing of the presentation layer and reducing overall cost of data maintenance and delivery.

Let us provide some analogue from telecom market. Shall we provide standards for SMS (SMPP) and MMS only or provide a standard framework for all messaging based services? Sure, we can have some benefits from the unified framework, but at the same time we may lose the flexibility. Guarantee, that such a standard model will cover all conceivable services is very low.

The key point is the simplicity and finally, time to market for new applications. Note, that telecom projects could be heavily affected by standard problems due to high diversity in the devices and use case models. Not taking into account the interests of developers to create standards, we risk facing a parallel existence of the standard model and the actually utilized the existing approaches.

References

[1] Namiot, D., & Sneps-Sneppe, M. (2014, June). On software standards for smartcities: API or DPI. In ITU Kaleidoscope Academic Conference: Living in a converged world-Impossible without standards?, Proceedings of the 2014 (pp. 169–174). IEEE.

[2] Trappeniers, L., Feki, M. A., Kawsar, F., & Boussard, M. (2013). The internet of things: the next technological revolution. Computer, 46(2), 0024–25.

[3] Gubbi, Jayavardhana, et al. "Internet of Things (IoT): A vision, architectural elements, and future directions." Future Generation Computer Systems 29.7 (2013): 1645–1660.

[4] Sneps-Sneppe, M., and Namiot, D. (2012, April). About M2M standards and their possible extensions. In Future Internet Communications (BCFIC), 2012 2nd Baltic Congress on (pp. 187–193). IEEE. DOI: 10.1109/BCFIC.2012.6218001.

[5] Egyedi, Tineke M. "Standard-compliant, but incompatible?!." Computer Standards & Interfaces 29.6 (2007): 605–613.

[6] Hawkins, R., and Ballon, P. (2007). When standards become business models: reinterpreting "failure" in the standardization paradigm. Info, 9(5), 20–30.

[7] Schneps-Schneppe, M., Namiot, D., and Ustinov A. "A Telco Enabled Social Networking and Knowledge Sharing." International Journal of Open Information Technologies 1.6 (2013): 1–4.

[8] Persson P., Jung Y. Nokia sensor: from research to product //Proceedings of the 2005 conference on Designing for User eXperience. – AIGA: American Institute Of Graphic Arts, 2005. – p. 53.

[9] Smith P. Comparisons between Low Power Wireless Technologies //US Patent CS-213199-AN. – 2011.

[10] Bluetooth S. I. G. Specification of the Bluetooth System-Version 4.0 //2010. http://www.bluetooth.com. – 2010.

[11] Otte R. Low-Power Wireless Infrared Communications. – Springer-Verlag, 2010.

[12] Namiot, D., and M. Sneps-Sneppe. "On the analysis of statistics of mobile visitors." Automatic Control and Computer Sciences 48.3 (2014): 150–158.

[13] Pongpaichet, S., Singh, V. K., Jain, R., & Pentland, A. S. (2013, October). Situation Fencing: Making Geo-Fencing Personal and Dynamic. In Proceedings of the 1st ACM international workshop on Personal data meets distributed multimedia (pp. 3–10). ACM.

[14] Nitro Mobile http://blog.nitromobilesolutions.com/category/technology/ibeacons/ Retrieved: Nov, 2014–11–18

[15] Silva, S., et al. (2012). A Bluetooth Approach to Diabetes Sensing on Ambient Assisted Living systems. Procedia Computer Science, 14, 181–188

[16] Rate, B., Rate, E. D., LE, L. E., Protocol, A., & Profile, G. A. Introduction to Bluetooth Technology.

[17] Android BLE http://developer.android.com/guide/topics/connectivity/bluetooth-le.html Retrieved Nov, 2014.

[18] Nahavandipoor, Vandad. iOS 8 Swift Programming Cookbook: Solutions & Examples for iOS Apps. "O'Reilly Media, Inc.", 2014.

[19] Usländer, Thomas, et al. "The future internet enablement of the environment information space." Environmental Software Systems. Fostering Information Sharing. Springer Berlin Heidelberg, 2013. 109–120.

[20] Moltchanov, B., & Rocha, O. R. (2014, April). Generic enablers concept and two implementations for European Future Internet test-bed.

In Computing, Management and Telecommunications (ComManTel), 2014 International Conference on (pp. 304–308). IEEE.

[21] Future Internet Public Private Partnership. Outcomes, achievements, and outlook. Phase 1, Final Report, European Commission, Sept 2013.

[22] FI-WARE Wiki https://forge.fi-ware.org/plugins/mediawiki/wiki/fiware/index.php/FI-WARE_Internet_of_Things_(IoT)_Services_Enablement Retrieved: Nov, 2014

[23] Namiot, D., & Sneps-Sneppe, M. (2014). On Micro-services Architecture. International Journal of Open Information Technologies, 2(9), 24–27.

[24] Sneps-Sneppe, M., and Namiot, D. "M2M Applications and Open API: What Could Be Next?." Internet of Things, Smart Spaces, and Next Generation Networking. Springer Berlin Heidelberg, 2012. pp. 429–439.

[25] Namiot, Dmitry, and Manfred Sneps-Sneppe. "Geofence and Network Proximity." Internet of Things, Smart Spaces, and Next Generation Networking. Springer Berlin Heidelberg, 2013. pp. 117–127. DOI: 10.1007/978-3-642-40316-3_11

[26] Fuhrhop, C., Lyle, J., and Faily, S. (2012, April). The webinos project. In Proceedings of the 21st international conference companion on World Wide Web (pp. 259–262). ACM.

[27] Arsénio, Artur, et al. "Internet of Intelligent Things: Bringing Artificial Intelligence into Things and Communication Networks." Inter-cooperative Collective Intelligence: Techniques and Applications. Springer Berlin Heidelberg, 2014. 1–37.

[28] Hunkeler, Urs, Hong Linh Truong, and Andy Stanford-Clark. "MQTT-S—A publish/subscribe protocol for Wireless Sensor Networks." Communication Systems Software and Middleware and Workshops, 2008. COMSWARE 2008. 3rd International Conference on. IEEE, 2008.

[29] P. T. Eugster, P. A. Felber, R. Guerraoui, and A.-M. Kernmarrec, The many faces of publish/subscribe, ACM Computing Surveys, vol. 35, no. 2, pp. 114–131, June 2003.

[30] Guinard, D., Trifa, V., & Wilde, E. (2010, November). A resource oriented architecture for the web of things. In Internet of Things (IOT), 2010 (pp. 1–8). IEEE.

[31] Namiot, D., & Sneps-Sneppe, M. (2014). On Micro-services Architecture. International Journal of Open Information Technologies, 2(9), 24–27.

[32] Bahrami, M., & Singhal, M. (2015). The Role of Cloud Computing Architecture in Big Data. In Information Granularity, Big Data, and Computational Intelligence (pp. 275–295). Springer International Publishing.

Biographies

D. Namiot is a senior researcher at the Faculty of Computational Mathematics and Cybernetics Lomonosov Moscow State University. Dr. Dmitry Namiot received a Ph.D. degree in Computer Science of the Lomonosov Moscow State University for his work in the area of artificial intelligence. His research activity is in the field of mobile computing, context-aware programming, Smart Cities applications and open interfaces for telecom applications. He is author or co-author of over 90 journals and conferences articles and 4 books. Dr. Namiot won several awards for mobile development on 3GSM World contests, as well as several international Java Developers challenges awards. He is the editor of International Journal of Open Information Technologies (INJOIT).

M. Sneps-Sneppe is Leading Researcher at the Ventspils University College, Latvia. His research focus is on telecommunications software development. His hobby relates to Russian history. He is (co-)author of 11 books and many papers.

An Open Source Real-Time Data Portal

Sudesh Lutchman and Patrick Hosein

Department of Computer Science, The University of the West Indies, Trinidad
E-mail: sudesh.lutchman@gmail.com, patrick.hosein@sta.uwi.edu

Received: 14 December 2014;
Accepted: 9 March 2015

Abstract

In the coming years there will be a rapid growth in the generation and processing of data from multiple sources. These sources, sometimes referred to as the Internet of Things (IoT), will consist of devices such as sensors, monitors and smartphones. This trend has been made possible because of (a) the adoption of IPv6 (allowing for such devices to be addressable), (b) enhancements in cellular technologies (in particular 5G) to better support short infrequent bursts of data transmissions and (c) the introduction of smartphones with a wide range of sensors and (d) the availability of easily deployed, low cost, low energy sensors (helped by the introduction of Raspberry Pi and Arduino processors). The real-time data generated by these devices can serve be used in several ways including being mined for important information and correlations, as well as use in web and mobile applications. This data, however, is often either proprietary or in a format that is not easily usable by real time applications. Furthermore, if an application developer needs to integrate (mash-up) data from multiple sources it is quite burdensome to develop interfaces. The present platforms for open data repositories are designed for static non real-time data. We propose a design and provide specifications for a platform that will accept real-time open data from multiple sources in multiple formats and make the data accessible in a standard format at user specified frequencies. The data is held for a finite period of time and digests made so that developers can also make use of historical data.

Journal of ICT, Vol. 2, 289–302.
doi: 10.13052/jicts2245-800X.235

Keywords: Open data, Real-time systems, Sensor networks, Data repository, Internet of Things, IoT, 5G.

1 Introduction

Open data, as defined by the Open Knowledge Foundation, is data or content that is free to use, reuse, and redistribute subject only at most, to the requirement to attribute and/or share-alike [4]. The World Bank further defines openness as a combination of being both technically and legally open [2]. Many Governments worldwide are currently going through an open data revolution. The focus is currently on increasing transparency [10] such as in the case of the United States of America with its Open Government Directive [5] and the European Union's Public Sector Information (PSI) directive [3]. As a result, many data portals and open data platforms have sprung up to accommodate the needs of these initiatives [7]. Examples include data.gov, data.gov.uk and data.tt. The World Bank has also been actively working with various countries and informing the Governments of the advantages of making their data open and available to the public.

These data platforms, such as CKAN (Comprehensive Knowledge Archive Network) and Junar, work well for processed, aggregated and summarized data but fall short of providing any efficient means of supporting raw real-time data. Furthermore, many institutions that deal with real-time data do so using proprietary methods. For example, CitySDK [1] is an open data portal that caters for a city's data but has a separate module specifically designed for dealing with real- time transport data. This makes it extremely difficult, and most times impossible, to use multiple real-time data sources in a collaborative effort. The lack of a proper system designed for real-time data also discourages many people from publishing much needed real-time data even if they are willing to provide it to the public. Platforms have been created to use real-time data such as the systems designed in [11] and [9]. These platforms manipulate the data to create Linked Data. Linked Data, however, simply is not able to guarantee the open data principle of reuse [12]. For geo-spatial data, the platform of choice is GeoNode but this does not support real-time sources.

The ability for almost anyone to develop web and mobile applications with minimal training has resulted in an explosion of these applications, especially for the Android and IOS smartphone markets. Some of these applications depend on real-time information and so developers need to work closely with the providers of the required data in order to ensure compatibility and also to come to an agreement on usage policies. If an application requires data from several data sources then the time, complexity and cost of development can grow rapidly. We propose a platform that would significantly reduce this burden by allowing the application developer to focus on the application and simply assume that the required data can be obtained in a standard format at whichever frequency is desired. Note that we assume that both content providers and content consumers will work according to some agreed upon set of policies and licenses.

In the next section we provide an overview of the currently available platforms that were developed for open data. We then compare them with our proposed platform which we call the Real-Time Open Data repository (RTOD) platform. Next we present the requirements for a real-time repository and indicate how these are incorporated in our proposal. In particular, we discuss how received data is processed. If the data is requested faster than it is collected then extrapolation techniques are used to provide data when it is needed even if a new value has not been received. Similarly if data is requested at a slower rate than what is collected then filtering (e.g. simply skipping some samples) can be used. Finally we provide some more implementation details of how application developers can access data via standard APIs (Application Programming Interface) and describe some Use Cases.

2 Overview of Current Platforms

In this section we provide a brief overview of the leading open data platforms, CKAN, Junar, GeoNode and the SensorMasher system designed in [9]. We follow the approach used in [7] to compare these platforms and consider the properties that they considered. We focus on the ease of uploading, downloading, accessing and using data. Although the GeoNode platform is Open Source and heavily utilized for the storage of geo-spatial data (e.g., GIS data) it does not satisfy all requirements of an Open Data platform and so is not included in our comparison table provided in Table 1.

2.1 CKAN

As described on their website ckan.org, "CKAN is a powerful data management system that makes data accessible by providing tools to streamline publishing, sharing, finding and using data. CKAN is aimed at data publishers (national and regional governments, companies and organizations) wanting to make their data open and available." CKAN is an open source open data

Table 1 Platform comparison

		CKAN	Junar	SensorMasher	RTOD
General Use	Supported Languages	Multiple	English and Spanish	English	English
	Registration	to upload data	to upload data	to upload data	to upload data
Metadata	Type	Contextual and detailed	Not very contextual and detailed	NA	Contextual and detailed
	Structure	Machine readable	Machine readable	NA	Machine readable
Open Data	Uploading	Registered users	Registered users	Registered users	Registered Users
	Target	Governments	Governments and businesses	Academia	Public
	Value Focus	data economy, innovation, transparency, collaboration	data economy, innovation, transparency, collaboration	data mashups	Real-Time Data
	Type	Multiple	Multiple	Multiple	Multiple
	Level	Mainly national	Mainly national	Low level sensor data	Any
	Format	CSV, XLS, PDF, etc.	CSV, XLS, PDF, etc.	RDF	Multiple Formats
Finding Data	Searching	Keyword search	Keyword search	Location search only	Keyword Search
	RESTful API	Yes	Yes	No	Yes
	Interfaces	Simple to use	Simple to use	Complicated controls	Simple to use
	Navigation	Clear	Clear	Not very clear	Clear
Using Data	Cost	Free to use	Fee	Free to use	Free to use
	Download	Metadata and data	Data only	Data in RDF format	Metadata and data
	Format tools	XLS, CSV to JSON	XLS, CSV to JSON	mashup option	Yes
	Collaboration	No	No	No	Yes
	Version control	Yes	No	No	Yes
	Visual data display	tables, maps and charts	tables and charts	line charts, limited maps and raw data	Real-Time graphs

platform streamlined for data publishers whether private or Government. CKAN allows registered users to upload data in many different formats including CSV, XLS and JSON. Each resource belongs to a parent dataset which contains metadata for the data. CKAN allows for entire datasets or single resources to be downloaded via the client web interface or through a semi-RESTful API. Data for select formats such as CSV and XLS can be retrieved as JSON, displayed as charts or even shown as tabular data. CKAN also has a geospatial component which allows for visualizing geospatial data on a map. CKAN has turned out to be the most popular Open Data platform and is used by many Governments worldwide (e.g., data.gov in the United States of America and data.gov.uk in the United Kingdom).

2.2 Junar

According to their web site www.junar.com "Junar provides everything you need to open data with confidence. Its built for massive scale and supports local and global data-driven organizations. With Junar, you can easily choose what data to collect, how you want to present it, and when it should be made publicly available. You can also determine which datasets are made available to the public and which datasets are available only for your internal use. Think of it as a next generation data management system for sharing information."

Junar is a specialized data platform geared towards governments and organizations at a commercial price which varies based on the size of establishment. Junar provides similar services as does CKAN such as the uploading of data in many different formats. Junar's service, being more tailored towards business needs, allows a user to enhance uploaded data by making custom charts. It also incorporates different analytic features for business and government use. Data in Junar can be downloaded via the web interface or programmatically retrieved through HTTP requests.

2.3 SensorMasher

According to the web site sensormasher.sourceforge.net, "SensorMasher is a platform which makes sensor data available following the linked open data principle and enables the seamless integration of such data into mashups." According to Wikipedia, "In computing, linked data (often capitalized as Linked Data) describes a method of publishing structured data so that it can be interlinked and become more useful. It builds upon standard Web technologies such as HTTP, RDF and URIs, but rather than using them to serve web pages for human readers, it extends them to share information in a way that can be read automatically by computers. This enables data from different sources to be connected and queried."

The SensorMasher system is quite different from both CKAN and Junar. SensorMasher gets its data through the Linked Open Data cloud in streams using the RDF format. It also does not provide any API to programmatically download the data. The data has to be manually downloaded through the web interface. SensorMasher allows a user to create interesting mashups between different streams of sensor data but the process is very difficult and not user friendly. The web interface, though unfriendly, does provide some data visualizations through maps and charts. The approach taken in [8] is also based on the RDF format.

2.4 GeoNode

According to their site geonode.org, "GeoNode is a web- based application and platform for developing geospatial information systems (GIS) and for deploying spatial data infrastructures (SDI). It is designed to be extended and modified, and can be integrated into existing platforms." GeoNode is the most popular Open Source geospatial content management system. GeoNode is used through a web application. Using the web application users can upload different types of geospatial data as layers. This data is made public through the very same web application which allows viewing the data by wrapping the layers over maps. Despite GeoNode's ability to upload, publish, download and view data, it was excluded from the open data platform comparisons because the system does not allow programmatic publishing or downloading of data. The system does incorporate some internal APIs that deal with customizing the system but there are no public APIs for manipulating data.

The aforementioned platforms, while providing helpful services, are not suitable candidates for a real-time open data repository. CKAN and Junar both target governments and business organizations, as a result they cater directly for summarized data in document form as opposed to real-time data which is granular and updated more frequently. The SensorMasher system uses real-time data but in the form of Linked Data from the Linked Open Data cloud and therefore it is not raw real-time data.

3 Specifications for Proposed Platform

The system architecture consists of three components (see Figure 1), the data providers (which are the sources of data), the Real-Time Open Data Repository (RTOD) (for data collection and dissemination) and the data consumers

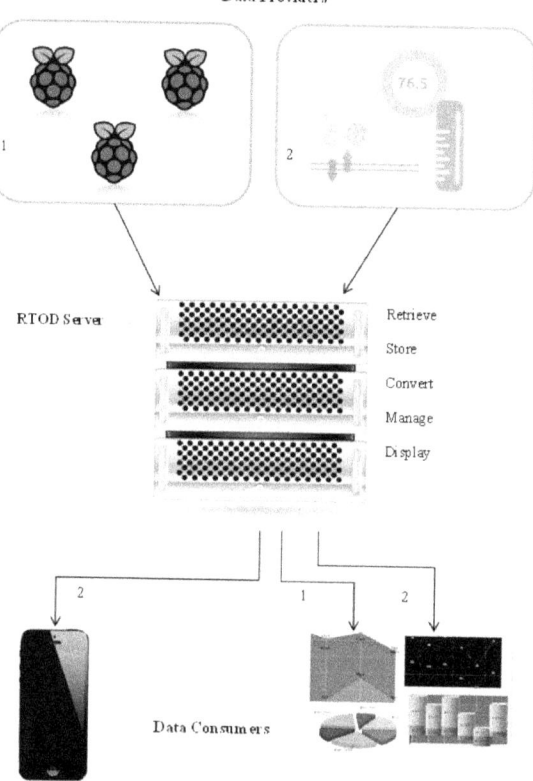

Figure 1 Platform architecture

(these are the developers who develop mobile and web applications for the end user). Hence we need to specify the interfaces between the data sources and RTOD, the data consumers and RTOD and the RTOD platform. In this section we provide an overview of this design. In the next section we provide more details of the software architecture of RTOD.

3.1 Data Providers

Each data provider may provide multiple streams of data. The provider must first register on the RTOD platform and also register each stream that will be made available. Details such as format of the data, frequency of generation, metadata for the streamed data and other pertinent data must be submitted as well as acceptance of the data policies of the RTOD. For each stream an appropriate license must be chosen by the provider under which the data is made available (e.g, the Creative Commons License which is specified at creativecommons.org). The RTOD platform supports various APIs for receiving data including the RESTful API. The data provider will also need to specify which protocol will be used. In summary the data provider must:

1. Accept and follow the policies and guidelines specified on the RTOD platform.
2. Register for an account on the RTOD platform.
3. Register each source of data that will be provided with details (metadata, frequency, protocol for uploading, data format and precision).
4. Begin uploading of periodic data to RTOD.

3.2 The RTOD Server

The RTOD server is used to collect data, process the data and allow others to retrieve the data. Its requirements, which are based on our earlier discussions, are as follows:

1. Provide a registration interface for data providers and allow for the registration and authorization of streams
2. Provide a registration interface for data consumers
3. Be able to simultaneously receive multiple data streams from multiple data sources in real time
4. Be able to store and manage data and metadata for each received stream
5. Be able to process (i.e., filter, extrapolate, convert, digest etc.) data from each stream
6. Be able to serve data streams to data consumers at the requested frequency (within limits)
7. Be able to display the stored data in various formats (tabular, graphical etc.) and allow users to manually download data
8. Provide search capabilities
9. Manage data and periodically flush dated data
10. Allow registered users to maintain their accounts (e.g., data providers can add/drop/modify data streams, data consumers can add/drop/modify requests for data and regular users can modify details of their accounts)

3.3 Data Consumers

The data consumers are the application developers who use the provided data in their mobile and web applications. Although they are called data consumers, the real consumers are the users of the applications which are the end users.

Application developers will be able to access the data via the standard RESTful API. This allows the developer to choose the rate at which data is retrieved based on the requirements of their application. The developer will be provided limits on the rate at which the specific dataset can be retrieved as well as the number of simultaneous streams that can be accessed so as to maintain acceptable performance levels for all developers. Data consumers must agree with the data policies specified on the RTOD platform. Note that these data consumers will develop user friendly applications for use by the general public who need not understand technical details such as APIs.

4 Implementation Details

The server was developed with the Python language using the object oriented programming paradigm along with various design patterns. The server uses the Waitress Web server because it has excellent performance and is also platform independent. All user tasks fall under a specific category e.g. an account task or dataset task. Each category is represented as a Service Class. A service is attached to a URL which gives users access to them. Each service class defines a function for every HTTP method that it supports, thereby allowing many actions to be represented by one URL. These service classes and associated URLs makeup the request handler (RH) component of the server.

Requests, that require querying the database, are passed over to the database handler (DH) from the RH. Every table in the database maps to a Model Class in the database handler. The DH then uses instantiations of these classes to communicate with the database via database sessions. In order to allow multiple concurrent users to perform database dependent tasks simultaneously a registry design pattern is used to allocate a database session to each client.

The server contains a views module that allows users to GET the client application. The client application is built on the MV* architecture using both HTML and JavaScript. The application itself is a single page application that uses its own Router class to determine which view to present. Views are populated using a combination of static data and dynamic data contained by the clients models. The clients models are similar to their server side brothers and are populated with data from AJAX calls to the RH, which are forwarded to the DH whose response is further forwarded to the client via the RH.

The client applications feed view provides real-time updates of the data stream that is currently being viewed, which are displayed on line graphs.

In order to avoid costly polling of the server both the client and server implement sockets. When a client goes to a feed view it sends a GET request to the socket service running on the server. When the socket service receives a socket connection request to the feed socket service the connection is established and added to the namespace. Once the client has established a connection it requests access to a given feed and the server then adds it to a list of sockets for that feed. The client application listens on the socket through non-blocking code for updates from the server. Each feed being viewed has its own socket connection. When the RH receives a POST request to a feed and the DH successfully inserts the data the RH passes the data to the socket service which then emits the data to all sockets associated with the given feed.

When a client navigates away from a feed the socket is disconnected and the socket service removes the socket from its lists. The socket service is micro-threaded allowing multiple sockets to be managed concurrently. See Figure 2 for a depiction of the software architecture.

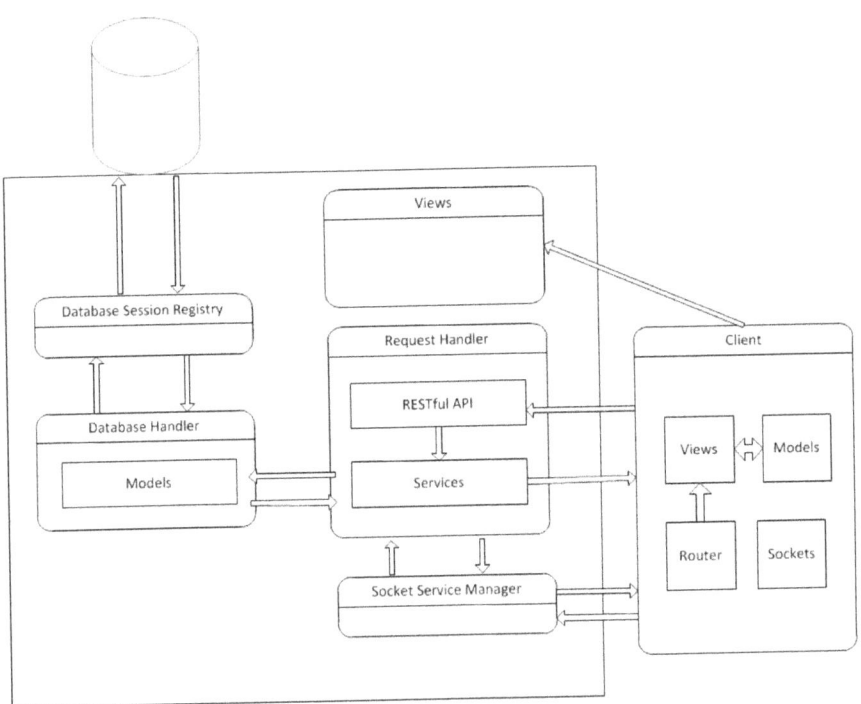

Figure 2   Software architecture

5 Sample Use Case

In this section we present a simple use case to illustrate usage of the RTOD platform. This simple example is just for illustrative purposes. The data source machine feeds its CPU utilization information to the RTOD server. This information is then retrieved by a third computer which then displays the data in graphical form. This example, although simple, illustrates the three components of the system. One can see that additional sources can be added and multiple applications using multiple streams can be used by simply following the process illustrated in this example.

Figure 3 provides a simple application for displaying the retrieved data (i.e. the CPU utilization of the data source) on the client (the data consumer).

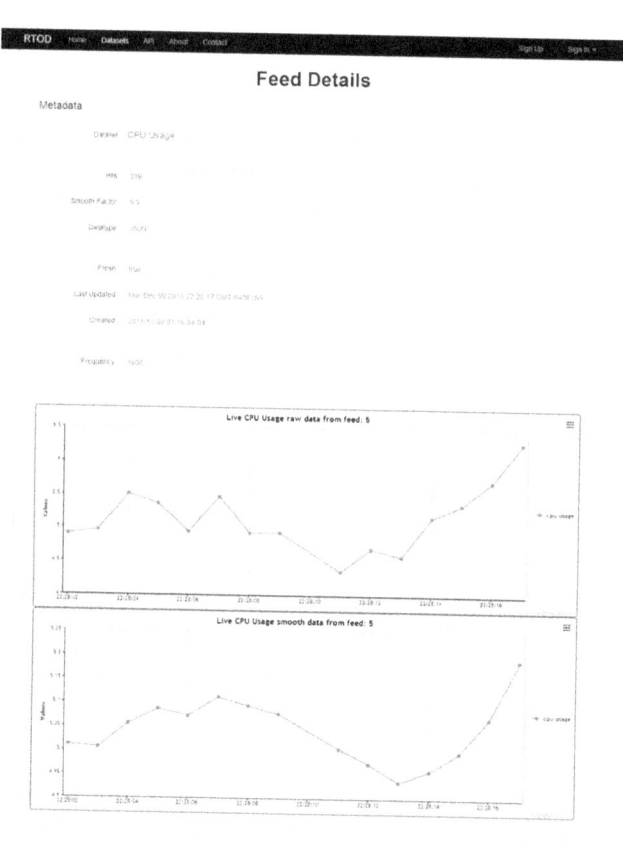

Figure 3 Sample use case

Note that the feed also provides metadata for the source such as the last updated time, data format and freshness of the data. In this particular case the client requested, in addition to the original feed, an exponentially smoothed version with a filter constant of $\alpha = 0.9$. The smoothed samples s_n are computed from the raw samples r_n as follows:

$$s_n = \alpha s_{n-1} + (1 - \alpha)r_n \qquad 0 < \alpha < 1 \qquad (1)$$

Both versions of the feed are displayed graphically and is updated in real time. The top graph contains the raw data while the bottom one contains the filtered version.

6 Conclusion

In this paper we provided requirements and specifications for an open realtime data repository. We use JSON as the data format as opposed to RDF since it allows for faster transfer and parsing of data. The temporal database is designed to handle a heavy load of incoming real-time data while providing multiple raw and/or processed streams to multiple data consumers (mobile and web applications). The design provides functionality that combines the requirements for open data, real-time data and an open data platform.

References

[1] About citysdk linked open data distribution api. available: http://citysdk .waag.org/about.CitySDK.

[2] The world bank. open data essentials, available: http://data.worldbank .org/about/open-government-data-toolkit/knowledge-repository. The World Bank.

[3] Directive 2003/98/ec of the european parliament and of the council of 17 November 2003 on the re-use of public sector information. available: http://ec.europa.eu/informationsociety/policy/psi/rules/eu/indexen.htm. European Commission, 2003.

[4] Definition of open data and content. available: http://opendefinition.org/. Open Knowledge Foundation Network, 2004.

[5] Memorandum for the heads of executive departments and agencies: Open government directive. The U.S. White House House, 2009.

[6] M. Janssen A. Zuiderwijk and K. Jeffery. Towards an e-infrastructure to support the provision and use of open data. In Proc. Conference for

e-Democracy and Open Government, Krems an der Donau, Austria, 2013.

[7] M. Janssen A. Zuiderwijk and A. Parnia. The complementarity of open data infrastructures: An analysis of functionalities. In *Proc. 14th Annual International Conference on Digital Government Research,* 2013.

[8] F. Corno and F. Razzak. Publishing lo(d)d: Linked open (dynamic) data for smart sensing and measuring environments. In *Proc. 3rd International Conference on Ambient Systems, Networks and Technologies (ANT)/9th International Conference on Mobile Web Information Systems (MobiWIS), Canada,* 2012.

[9] D. Le-Phuoc and M. Hauswirth. Linked open data in sensor data mashups. Digital Enterprise Research Institute, National University of Ireland, Galway, Ireland, 2009.

[10] K. OHara. Transparency, open data and trust in government: Shaping the infosphere. In *Proc. 3rd Annual ACM Web Science Conf.,* 2012.

[11] H. Patni. and C. Henson. Linked sensor data. In *Proc. Collaborative Technologies and Systems (CTS) Symposium,* 2010.

[12] D. De Roure P. Missier J. Ainsworth J. Bhagat P. Couch D. Cruickshank M. Delderfield I. Dunlop M. Gamble D. Michaelides S. Owen D. Newman S. Sufi S. Bechhofer, I. Buchan and C. Goble. Why linked data is not enough for scientists. *Future Generation Computer Systems,* 29:599–611, 2013.

Biographies

Mr. S. Lutchman Sudesh attended the University of the West Indies where he obtained a BSc. and a MSc. in Computer Science. He was an ACM ICPC regional participant for 2012 and a coach for 2013. He has worked as an instructor at both the University of the West Indies as well as at the University of Trinidad and Tobago and has consulted as a software developer. His interests include open data, data mining, artificial intelligence and mobile application development.

Prof. P. Hosein. Patrick attended the Massachusetts Institute of Technology (MIT) where he obtained five degrees including a PhD in Electrical Engineering and Computer Science. He has worked at Bose Corporation, Bell Laboratories, AT & T Laboratories, Ericsson and Huawei. He has published extensively with over 75 refereed journal and conference publications. He holds 38 granted and 42 pending patents in the areas of telecommunications and wireless technologies. Patrick is presently the administrative and technical contact for the TT top level domain, CEO of TTNIC and a Professor of Computer Science at the University of the West Indies. His present areas of research include radio resource management, QoS and pricing for 5G cellular networks.

Author Index

303

Keywords Index

www.ingramcontent.com/pod-product-compliance
Lightning Source LLC
LaVergne TN
LVHW012332060326
832902LV00011B/1845

* 9 7 8 8 7 9 3 2 3 7 9 1 9 *